Was Mona Lisa created by physicochemical reactions alone?

Open your mind and use your logic…

Nickolas Dorfman, Ph.D.

iUniverse, Inc.
New York   Bloomington

Was Mona Lisa Created By Physicochemical Reactions Alone?
Open Your Mind and Use Your Logic...

iUniverse books may be ordered through booksellers or by contacting:

iUniverse
1663 Liberty Drive
Bloomington, IN 47403
www.iuniverse.com
1-800-Authors (1-800-288-4677)

Because of the dynamic nature of the Internet, any Web addresses or links contained in this book may have changed since publication and may no longer be valid. The views expressed in this work are solely those of the author and do not necessarily reflect the views of the publisher, and the publisher hereby disclaims any responsibility for them.

ISBN: 978-0-595-51843-2 (pbk)
ISBN: 978-0-595-62058-6 (ebk)

The front cover was created by Russian artist Ekaterina Shulzhenko

Printed in the United States of America

Contents

"Judge a man by his questions rather than by his answers"
Voltaire

"All truths are easy to understand once they are discovered; the point is to discover them."
Galileo Galilee

Introduction

Like all kids in the Soviet Russia, I was raised in a strictly atheistic environment. My middle class parents got an even stronger atheistic education. From a very early age during my kindergarten years, all my school years and then every year in the university, I was learning only materialistic viewpoints. Such severe brainwashing succeeded and I strongly believed that there is no God and any religion is an "enemy of people" and "poison for people's mind."

However, my early interest in biology somehow planted many questions in my mind; I was trying to find answers from my teachers and mentors. This search brought me to the department of biology at the Moscow State University. This is the university from which I graduated in 1965 and started my scientific research in the field of Tumor Immunology. At that time, I was still an atheist and religion was interesting to me only from historical and cultural viewpoints. Nevertheless, seeds planted in my mind during my childhood were growing and persistently tried to turn my mind to the dark side of science, where questions can be answered only by logic or by faith.

The human intellect was the most puzzling subject; I had a burning desire to get a clear understanding of what it is and how it works. While working in the field of science, which is mostly a materialistic environment, I was trying to analyze the human mind from the viewpoint of evolution and molecules, or in other words, rejecting everything that is beyond matter and energy. This brought me to a dead end with no exit.

As far as I remember, the most dramatic turn of my thinking happened when for the first time, I watched the wonderful TV program "60 Minutes" about blind musicians. I was trying to comprehend their incredible ability from the biological data available in science to date. At that point I saw a clear contradiction in my mind between science and logic. My second shock occurred when I met for the first time (not personally) Kim Peek and Daniel Tammet and learned about their unique abilities, described here in Chapter "Savant Syndrome" I fanatically started collecting any information I could find about Savant Syndrome. The collected data strongly reinforced my deviation from purely materialistic viewpoints. I still do not accept any religion without questioning and I do not consider myself a religious person. I do not attend any church, do not pray and never ask God about anything. All these things still do not make any sense to me. However, my new way of thinking has brought me to the belief that the human mind cannot be (and, most likely, is not) the only source of intelligence in the universe. There is something else; something that we still know nothing about; something that is beyond matter and energy.

Like anyone else, I cannot give you any scientific proof of God's existence, but I can call your attention to those facts of life that cannot be explained at this point without admitting a certain power that is outside of our brains. People call this power by many different names – God, Light, Consciousness, Source, Presence, Infinite, Oneness, Universal Intelligence, Divine, Great Spirit, Creator, Love, etc. It does not matter what name people prefer as long as they realize that chemical reactions in our brains cannot be the only sources of intelligence and cannot give us that quality we call intellect.

In this book, I am trying to find answers to numerous questions about God and his influence on the human mind. Any discussion on these subjects can be done only if we use an open mind, logic, speculation and even fantasy. The alternative way is the way of atheists, who see only brain cells and biochemical reactions inside of them. Although their way can provide a lot of scientific information, it will never be able to answer a simple question: "Who created the Mona Lisa?"

In this book, I am not going to torment you with such New Age terms like "multidimensional creation," "permeable boundaries," "another dimension of reality" and many others. I did my best to write this book in a popular form, but life is a too complicated process and it is not always feasible to present it in simple terms. If you find some parts of this book as too heavy to digest, just skip them and read further. Scientific data, along with intriguing, but logical speculations and some fantasy, presented in this book, give you a lot to think about regardless of your religious or atheistic inclination.

Evolution, Creation or some Combination?

What is the origin of life on the Earth? What is a human being? What is an intellect? How does it work? These and many other questions come to the minds of every person who is capable of thinking. Although there is not an accurate answer to these questions, man finds a certain satisfaction in acceptance of one or another theory, hypothesis or religious explanation. There are many theories in this field. The majority of them fall into one of two main categories - evolutionary or creational. The followers of each of these two categories continue their endless arguments about whether life was created by God or by evolution.

There are other hypotheses that cannot fit into one of these two categories. On the Earth, according to one of these hypotheses, life was brought from a cosmos in the form of some primitive organisms, like bacteria, or brought by intellectual creatures from other planets. Such hypotheses practically explain nothing about the origin of life, but support the evolutionary theory, assuming that primary life on Earth was not created, but was brought to our planet. Is it a better or more scientific explanation, than that which states that everything is created by God?! Religious theories are more attractive to the majority of people because, by accepting and believing in a certain religion, there is no need to ask yourself the question regarding how human beings and all the incredible varieties of other organisms have appeared on this planet. It is easier to live with a faith, since you know that your entire life and everything around you is under the control of God, hence no matter what you want or don't want - God knows best and controls everything. Having a faith means you do not need to think about how man was created and what will happen to him after death. Everything is explained in layman's terms; there is no need to think, no need to seek for the truth - you just have to believe. And as the Russian adage states, "the less you know, the better you sleep"; and this is good for your health.

This kind of truth however cannot satisfy everyone. A minority of people cannot accept faith without questioning. Their brains are designed in such way that they can believe in things that they can see, smell, touch, hear, or at least measure. For these people evolutionary theories are more attractive. I doubt that anyone, who knows at least the basics of evolution, can deny that a very important role of it involves the development of endless variations of life forms. However, to believe that evolution alone is responsible for the creation of such an incredibly sophisticated machine, like the human being, is as difficult to believe as that a car can be assembled by Mother-Nature without a man's hand. Imagine that a car would emerge just because of random occasional movements of chemical elements and their interaction with other molecules of an inanimate nature. But that is exactly how supporters of the evolutionary theory explain the origin of life. Concisely, it happened like this:

Once upon a time, one molecule, scudding in a primordial ocean, suddenly and **absolutely accidentally** met another cute molecule. They decided to bind to each other and scud together, and found that they acquired a new property, which gave them a better chance to survive in that rough and unfriendly primordial ocean. During the next billions of years, these molecules met many other attractive molecules (of course, **absolutely accidentally**) and combined with many of them, creating a big conglomerate of molecules. This conglomerate again **absolutely accidentally** discovered that it can create similar little baby conglomerates by simple division and that is how a primordial organic complex or protein or nucleic acid was created. They then **absolutely accidentally** found each other in the primordial ocean, combined and created a small syndicate that acquired many new and very useful properties for survival and we now call it a "cell."

Being educated in a spirit of materialism and fully denying the existence of God, I could accept this "fairy tale," but only with one essential reservation. Namely, **accidental and random combinations of molecules can neither create a car nor a human being or even a simple one-celled organism without a certain "Directing Power" that can transform originally inanimate matter into numerous**

forms of life. Evolution is just a tool that allows endless transformation of one kind of life form into others in order to adapt them to dynamic changes in the surrounding nature. But could evolution alone create a simple form of life? If we believe so, then I do not see much difference between believing in this theory or believing in God. Both faiths do not have enough ground to explain how it all happened in the beginning. Atheism is just another religion that uses evolution as the main argument of their religion, but still cannot explain numerous facts of life, its origin and numerous biological phenomena.

If evolution alone could not turn inanimate matter into life forms, then what is that Directing Power that was able to do so and then improve it to such incredible sophistication as we can see in a human being? Can we call this power "God?" It can be called anything you want and it does not change a thing if you do not see God as a human-like creature, sitting somewhere in the clouds, watching over every livable creature on the Earth and keeping a certain order on this planet. Religious dogmas have some wisdom, but also contain a lot of fantasies, which are just a reflection of people's folklore, their culture, and their lives.

What is a WISH or a THOUGHT?

I am writing these lines, flying from Washington to San-Diego (year 1989) and wondering: - "How could that relatively small collection of proteins and other organic elements in our brain create, from inanimate matter, such a huge machine as the Boeing, and propel it against gravity into the air? How have all those molecules of our brain, which represent just different combinations of five basic elements (carbon, oxygen, hydrogen, nitrogen and sulphur) managed to work together in such incredible cooperation to lead to the development of inanimate matter into something flying, running, swimming and even thinking?"

Atheists assert that brainwork is a complex physicochemical process. Who would argue?! But are those physicochemical reactions solely responsible for the creation of the Mona Lisa by Leonardo da Vinci, or the Pieta by Michelangelo, or heavenly music by Mozart? Are they solely responsible for building a space station, for sending man to the Moon, for building sophisticated computers and for "emulating God" by playing with codes of life?

Here, my hand is writing some scrawls on paper, transferring some physicochemical processes from my brain first into muscles of my hand and then into my pen, making the inanimate matter (my pen) record one "crazy" thought after another, which have flashed through my mind. But what is a **THOUGHT?** Physicochemical process? Does it mean that again one molecule in the ocean of my brain found another molecule (**absolutely accidentally**), combined, gave a signal to a third molecule, which in turn did something else and, as a result of this chain reaction, a "genius" thought was born? Then this thought, in incredibly small fractions of a second, ran in the form of electrons along wired nerves to muscular cells and made them contract in such a precise order which allowed this whole instant process to be transferred on paper in the form of a recorded thought.

Such a simplified explanation of the brain's work might be quite acceptable, but only for some physiological processes. For example, a man is hungry. This situation causes too much acid to accumulate in his stomach, which creates an unpleasant environment for the epithelial cells of internal tissues. In response to environmental fluctuations they somehow change their chemical composition, which is recognized somehow by nearby nerves and numerous signals are transferred along the nerves to corresponding parts of our brain. Excitation of those areas in the brain gives us a sense of hunger and we run to kitchen, restaurant or Mommy.

Equally, other physiological processes could be explained in a similar way. However, it is not clear how to explain by merely physicochemical processes our actions, which are not caused by hunger, fatigue, temperature, etc. For example, I raise my hand. Nothing and nobody makes me to raise my hand. I can do it without any obvious reason and at any time whenever I want. This means that all that is required for my brain to transfer corresponding signals to my muscles and raise my hand is just my wish. But what is a **WISH?** Again, a physicochemical process? Again one molecule **accidentally** found another one? But if my wish can be repeated as many times as I want, how can this process possibly be **accidental**?! On the contrary, there must again be some kind of "Directing Power" that cannot be just an accidental change in chemical composition of a neuron or its excitation, but something that itself creates such changes and then leads them to the realization of a **wish** or a **thought**. Thus, in the brain's work, simple logic makes us think about a certain role of the Directing Power rather than an accidental combinations of chemical molecules generating electromagnetic signals. What is this Power?! How does it work? Where is its source?

Brain signals

Numerous experiments in brain behavioral studies have shown that our brain consists of many separate areas responsible for certain feelings or actions. Dr. Wilder Penfield (1891-1976) worked as a brain surgeon. Operating on epileptic patients, he discovered that by applying weak electrical currents to a patient's brain, he could induce movements of different muscles. Moreover, since these patients were anesthetized only locally during the operation, they could tell the doctor what they felt at the moments of when electrical stimulation applied to any particular part of the brain. With this kind of experiment, he was able to map different activities or feelings in different areas of brain. He found that in some brain areas he could artificially induce broad spectrums of memories, while different spots of the motor cortex generated movements of only those muscles that are assigned somehow to those individual brain spots. In one experiment his patient clearly heard a familiar song though it was not played in the operating room. In other experiments they expressed a feeling of a particular taste, hunger, fear, etc. His pioneering work has discovered that temporal lobes of the brain are responsible for the storage of memories, and the motor cortex is responsible mostly for muscle movements. Of particular interest are his experiments that showed that each memory has a physically identified spot in the brain. Repeated application of electrical stimulation to the same spot induced sounds of the same song or the same event from the patient's childhood, etc. Such physical storage units, he called "engrams."

Experiments with electrical stimulation of the brain areas were started even before Dr. Penfield. In the middle of the 19th century, Dr. Edward Hitzig and Gustav Fritsch worked in a homemade laboratory with dogs and showed that specific movements are controlled by specific brain areas.

At the present time we have detailed maps of brain areas controlling numerous specific functions. We also know that an epileptic seizure is a result of an electro-magnetic storm in the brain, which starts at one point and goes to other areas causing the contraction of many different muscles. However, we still know virtually nothing about why it happens. Is this just internal biochemical malfunction or does the brain start its excitation by a certain energy coming from the outside? If electrical stimulation of a particular brain area causes some chain reactions resulting in a certain action, can we assume by extrapolation that under normal conditions all actions of an organism begin from a certain excitation of a corresponding areas in our brain, caused by some signals, carried by electro-magnetic waves, which come to us from outside? Actually, that is exactly how it happens when waves of light (outside source, isn't it?) bring into our eyes information about things or actions around us. However, excitation of a certain area in the brain by visual information induces not just a fact of the presence of an object, but numerous other thoughts that somehow in a super fast dance create a final decision that determines the relationship of an organism to an object (e.g. to run toward or away).

Ron Hubbard, in his book "Dianetics" tried to explain it as the ability of the brain to memorize numerous pictures of life events that we experience or observe during our life. By some unexplained technique he calculated that the brain can store in memory 25 pictures per second. These pictures can be retrieved in any future event and influence our ability to make a decision or to act appropriately. Like anyone else at this level of modern science, Ron Hubbard could not explain how those pictures can be stored by the brain or instantly retrieved when needed.

Is it possible that virtually instant analysis of numerous variants of our relationship to a particular object is a result of random or accidental interaction of chemical elements in our brain? It seems absolutely unrealistic – doesn't it?! In the following examples I will show you that the brain can do amazing things without any pictures of a previous experience and without our consciousness. It is extremely difficult to explain these and numerous other biological phenomena without assuming a certain influence of a Directing Power that is capable of sorting an enormous number of variants in an unbelievably short time and selecting the most optimal for a given situation without any influence from our consciousness.

Are we human beings or computers?

The chemical composition of the brain is quite poor, but still it accumulates huge amounts of information. It would be relevant to note that the chemical composition of magnetic tape or a hard drive is even many times more simple, but still can store a lot of information. And when such information is transferred via electro-magnetic waves to a distant mechanical robot, like a spacecraft, it can make robots perform virtually any task, any movement, any measurement, etc. However, in such a process, a human hand is the primary (or maybe sub primary) source of signals that are first recorded, then transferred and transformed into actions. If any link in this chain is missing or malfunctioning, all those robots become just a combination of metal and plastic parts either completely dead or functioning incorrectly.

I do not exclude a possibility that we are all robots too, but just significantly more sophisticated. Our lives and all our actions completely depend upon information recorded in our brains. This information and its transformation into actions make the difference between a genius and an idiot, or between a crazy person and a normal one. Destroy this information or cut off its transfer to peripherals and the human robot will be turned into a dead combination of stinky organic substances or other robots will put him/her into a nut house.

The human brain works on the same principles as the brains of mechanical robots. Both of them have electronic memories; both of them can transfer this memory along wires or nerves to peripheral organs (legs and hands or printers and monitors), both of them can analyze information, releasing a certain decision. The most important fact is that both of them work on the basis of electro-magnetic interactions among all structures. Nobody, however, would argue that an inanimate computer has that Directing Power in the form of our hand working with the keyboard and sending signals into the computer's brain. But if the inanimate computer has our hand as a Directing Power, where does the Directing Power of live computer come from? From God? Or from chemical reactions inside of our neurons? Where is the source of that power? And is it wise and logical to deny this power just because we know nothing about it?

Too many questions - not so many answers. It seems to me that the Directing Power is a logical necessity for both live and inanimate computers. However, a live organism and especially a human being is a much more sophisticated machine than any modern computer. Such a machine can be self-programmed, self-improved, self-repaired and self-reproduced. Does this mean that live computers do indeed perform all those things by themselves without any influence from a Directing Power? That is hard to believe.

Many bright-minded scientists reject this idea just because there is not any evidence to support it. They are more comfortable with the theory of evolution and they are afraid of stepping into dark areas where there is no money for research and only numerous attacks from other scientists and politicians. We do not want to feel like we are puppets. We do not want any competition from any Directing Power, or God, or Intelligence Designer. We are the Masters of the Universe!! We are the only creatures of the Universe who can perform independent thinking, who have consciousness, who can use their independent intellect to concur with other planets and even to produce new life forms, which Mother-Nature did not have enough time to create!!!

Why are we so arrogant? Why do we close our eyes on a very logical perception of the fact that the Mona Lisa could not be created solely by physicochemical reactions in the brain? Why do we deny any possibility of an intelligence outside of our brain when we know perfectly well that our brain consists of the same basic elements as any other form of matter in the Universe? At this point I would like to show you a few remarkable examples of biology that cannot be explained by any postulate of modern science.

Blind musicians.

I am sure you have heard about blind musicians. I have personally met one of them in my life and will never forget what I saw. In case you have not had a chance to see one, I suggest that you contact the group of journalists at "60 Minutes" and ask them about the videotape they prepared about blind musicians. This group of the most talented and respectful journalists in American media have selected three remarkable examples of blind musicians of different ages and genders. All of them learned to play music at a very early age and mostly by themselves. One boy, for example, started playing Beethoven before he first said "Mommy." The majority of blind musicians are also mentally challenged. Yet they exhibit some musical abilities that other normal musicians do not have at all. In some experiments, a blind musician was exposed to a long piece from a sophisticated concerto, which he had never heard before. Immediately after listening, he was able to reproduce every note of this entire concerto with remarkable accuracy. Moreover, he could give several of his own improvisations of the same concerto. The same ability, which I call a real miracle, has been observed in many other blind musicians, who are victims of autism – a medical disorder of the development of the brain, which is called savant syndrome. When you see the performances of blind musical savants and if you have at least a very basic ability of logical thinking and a very basic knowledge of biology, you can be immediately convinced that these miracles just **cannot** be explained merely by physicochemical reactions. There must be something else!

It is obvious that the brain of a blind musician can somehow make an immediate recording of any piece of music and store it in the brain for a long time, even for life. Our brain is a very efficient recording device, but we still know nothing about how a quite limited variety of organic molecules in our brain are capable of recording enormous amounts of information. But this is only a small part of that miracle.

After recording, the brain of a blind musician is capable of making his hands and fingers (and legs too) dance in accurate coordination to reproduce the same piece of music or endless variations of the same music. What does modern science know about this? Practically nothing.

We know that every move of our hands, fingers and every muscle is a result of certain electrical signals generated in our brains. These signals are sent via neurons – wires to our muscles, that then are responsible for moving certain parts of our body in a strict order controlled totally by brain signals. As I mentioned before, a human being along with other live creatures is a sophisticated biological computer. A brain is our "hard drive" that probably has more gigabytes for storage of information than any modern mechanical computer. Our nerves are wires through which our "hard drive" communicates with peripherals – hands, fingers, legs, etc. As a result, our peripherals can create music, novels, art, this book and many other things that are always a result of a very complicated digestion of huge amounts of information constantly deposited into our brains throughout our entire lives.

The fact that we are biological computers does not surprise me. It is even obvious to me. The most intriguing question for me is: "Who is working the key board?"

Savant Syndrome.

We all learn from our childhood how to calculate. In elementary school we start downloading into our "hard drive" simple calculations. They help us later to perform more complex calculations. But when we need to calculate numbers over three or four digits, we find ourselves looking for a calculator to help us. However, there are some unique individuals who can do complex mathematical calculations with numbers consisting of numerous digits. They do it fast and without any calculator. One such phenomenon was brilliantly portrayed by Dustin Hoffman in the movie, "Rain Man." The Rain Man was not a Hollywood fantasy. The real Rain Man – Kim Peek – who was a prototype for the movie character, is the most unique savant living today. He was born with some brain abnormalities. A particularly interesting fact is that his brain does not have Corpus Callosum – the part of the brain that makes the connection between the left and right hemispheres. A few other parts are either missing or damaged. Kim is a living encyclopedia. He had memorized about 9,000 books and keeps in his memory encyclopedic knowledge of literature, history, music, sport, geography, all dates, names, events, etc. His ability to read is just unbelievable. His eyes are like two scanners. They scan two pages at the same time simultaneously in seconds and his brain memorizes immediately the contents of any text. He also has a unique calculating ability being able to name any day of the calendar many years back or forward. Having the unique abilities of a genius, he also has certain disabilities, which do not allow him to live without the constant supervision of his father.

Another unique giant among savants is Daniel Tammet. His brain abilities to calculate and to memorize are no less impressive. But he does not currently have any disabilities that are so common among savants. He lives a normal life in England and has his own business. In his childhood he had several epileptic seizures, which are probably responsible for his unique savant abilities. Like Kim Peek, he can easily name any day of the calendar many years back or forward. In one example of his quick calculation he was asked to raise 37 to the power of 4. The correct answer of 1,874,161 was given in less than one minute. In another episode he was asked to divide 13 by 97. The interviewer's computer calculated 32 decimal digits, but Daniel gave about 100 decimal places and could have given even more if he was not interrupted.

On March 2004, he set up a new European record by reciting from memory "Pi" number to 22,514 decimal places. He also has a remarkable ability to learn foreign languages. He fluently speaks ten languages. The difficult Icelandic language he learned in seven days. In his autobiography, "Born on a Blue Day," Daniel describes his life and his mathematical abilities. For me personally, the most interesting part of his book was his description of how he makes all those complex calculations without any computer or calculator. It turns out that his brain is not even trying to make all the calculations the way we do them. It all happens automatically and he just sees different colors, shapes and textures that create a certain landscape. The conscious part of his brain is just reading the sequence of those shapes and textures and assigning numbers to each of them. So, when he gives the right result of his calculation, he just retraces the landscape in his vision and reads the numbers from that picture. For a normal human being it is virtually impossible to imagine what he actually sees and how he reads those pictures in his mind. When other savants, who have similar extraordinary mathematical abilities, are asked how they do it, they usually respond: "I do not know. It just comes out of my mind." Here is how Daniel himself describes his unique abilities:

"My favorite kind of calculation is power multiplication, which means multiplying a number by itself a specified number of times. For example, the square of 72 is 72x72=5,184. Squares are always symmetrical shapes in my mind, which makes them especially beautiful to me. Multiplying the same number three times over is called cubing or "raising" to the third power. The cube, or third power, of 51 is equivalent to 51x51x51=132,651. I see each result of a power multiplication as a distinctive visual pattern in my head. As the sums and their results grow, the mental shapes and colors I experience become increasingly more complex. I see 37's fifth power – 37x37x37x37x37=69,343,957

– as a large circle composed of smaller circles running clockwise from the top around. When I divide one number by another, in my head I see a spiral rotating downwards in larger and larger loops, which seem to warp and curve. Different divisions produce different sizes of spirals with varying curves. Different tasks involve different shapes, and I also have various sensations or emotions for certain numbers."

Dr. Darold Treffert, a psychologist, who spent most of his career studying Savant Syndrome, has collected many remarkable examples of unique abilities of his patients. On the website of the Wisconsin Medical Society, a good collection of savants is presented with the description of their abilities and some video clips, which make your mind ponder – "How in the world is this possible?" In his extremely interesting book "Extraordinary People," Dr. Treffert describes many savants, whose skills, in my opinion, are the most convincing proof at the present time supporting the view that the human mind is something that cannot be reduced to merely genes and biochemical reactions. Here is one excerpt from this book, describing one savant – a blind man with "a faculty of calculating to a degree little short of marvelous":

"When he was asked how many grains of corn there would be in any one of 64 boxes, with 1 in the first, 2 in the second, 4 in the third, 8 in the fourth, and so on, he gave answers for the fourteenth (8,192), for the eighteenth (131,072) and the twenty-fourth (8,388,608) instantaneously, and he gave the figures for the forty-eighth box (140,737,488,355,328) in six seconds. He also gave the total in all 64 boxes correctly (18,446,744,073,709,551,616) in forty-five seconds."

Neurologist, Dr. Oliver Sacks, describes autistic twins in his book, "The Man Who Mistook His Wife for a Hat." These twins were severely autistic, that is, unable to take care for themselves, or to express themselves in normal ways of communication with other people. But their mathematical skills and especially calendrical calculations could not be called less than a miracle. They easily could give you the day of the week for any date over a 40,000 year span.

It is so obvious that their consciousness has nothing to do with this miraculous ability. They do not show any efforts in doing these calculations. Their brains just send electrical signals in a very special order to muscles of their tongues, lungs and vocal chords that provide sounds with information on a given task. The twins even do not understand what they are talking about, but the answer is always completely accurate.

What could be a more convincing evidence of the fact that our brain is only a tool for transmission of information that is released through our muscles in the form of a sound or handwriting? How is it possible to assume that trillions or more immediately released electrical signals from the brain in a very strict order can be a result of merely biochemical reactions in neurons? To believe in that, one must be a very naïve person with no logical thinking or as ignorant as all atheists and materialists.

We all know perfectly well that our brain sends constant signals to our heart and to the muscles of our lungs to perform correctly through our entire lives. Those signals are sent regardless of our consciousness, but we are not surprised with such autonomy of our brain performance. Why are we surprised then, if the same brain can do something else beyond our consciousness – like to work with musicians' fingers or to calculate complex numbers? Mechanical computers can do all the same things, but in the beginning of all those things are certain signals that we send when we are working with a keyboard. Who is then sending signals to our brain for its performance?

Incredible coordination.

On TV and in real life we have the opportunity to see huge flocks of birds and insects or shoals of fish. These concentrations of live creatures sometimes consist of thousands, millions or even billions (for insects) of individual independent organisms. All of them move separately but close to each other. Yet they do not collide and show remarkable synchronization of their movements. Entire flocks or shoals behave like one organism. If a flock has to turn right, each individual member of the flock turns in the same direction at the same moment. In the case of shoals of small fish, it is amazing to watch how millions of individual organisms change the direction of their movement almost every second without the creation of a disordered crowd. How do they do it? Maybe they have some kind of a Leader at the front of the line, who sends them commands like: "Hey, guys! We all turn left now, and now up, and now under at the angle of 35.27 degree to the right, and now down, but only 3.467 feet! You're your distance, breathe deeply, lunch is soon!"

We know now that each move of an individual organism depends on signals that are sent from the brain to the muscles. When a move of millions of individuals is strictly synchronized, it means that the brain of each organism sends the same signals at the same moment to the same muscles. In this example, millions of independent live computers create some kind of network. In order to comprehend how the same signal can appear in each "computer" at the same time, we have to assume two alternative situations. First, the signal is generated inside the brain in response to surrounding factors. Second, the signal is sent to all flock members at the same moment from outside. But then, who is sending it?

Synchronized Reproduction.

Coral represents a complex society of sea creatures, and coral reefs represent a huge accumulation of many different species of corals extending sometimes many miles. Although coral consists of hundreds or thousands of individual organisms, each of them lives independently and maintains a hermit type life style since they cannot move out of their houses. This kind of life style is supposed to restrict their communications with relatives living many miles from their side. Nevertheless the synchronization of their reproductive behavior regardless of distance is just amazing. Exactly one night (five nights in some other locations) after the day of the full moon and at a certain time between sunset and midnight all corals of a reef (no matter how long it is) start releasing millions of eggs and sperm into the water giving a great fiesta to fish and enormous pleasure to curious tourists. This process is called mass spawning and it happens once a year. It is one of the most amazing biological phenomena. Who gives them the signal to perform a synchronized reproduction? Each individual coral keeps its eggs and sperm in storage even if they are ready for conception, but releases them only when all other corals, even very distant ones, accept the same signal from outside for the synchronized reproduction.

Life is full of miracles. Just look around and you will see a lot of weird activities among live creatures, activities that nobody can explain in scientific terms. When we approach that border beyond which we know and understand virtually nothing, we use the term "Instinct." This term is a signature of our inability to explain the miracles of life. Look, for example, at a newborn kangaroo. Just a few centimeters of skinny bones make a long and very difficult, but unmistakable journey from the ground spot, where it was delivered to its safe place in its mother's pouch. How is this tiny cluster of organic substances aware of where to go and what to do? Instinct? Miracle? Inherited Genes? Or maybe it is just a tiny acceptor of signals from outside that help him move his naked body in the right direction to a new spot, where there is plenty of food, where it is very warm and extremely safe. Before I start speculating in the dark area of science, I would like to provide you with a little more undeniable science.

How do cells work?

Modern biologists have numerous sophisticated tools and techniques to study cells on the molecular level. Since the time of Drs. Crick and Watson, who showed us how genetic codes are designed in the molecular structure of DNA, we now know about virtually all the genes of the human genome and genomes of many animals. In a single experiment we can see which genes of a cell are actively working and which are in a dormant state in any particular stage of its differentiation or physiological stress. We can activate dormant genes and suppress active genes. Genes produce millions of important proteins and through them they regulate numerous physiological functions of any single cell and ultimately of an entire multi-cellular organism.

I am not going to present you with a biology lecture here; you should know the basics of biology from your school years. The point I would like to make here is that there is a surprising dramatic gap between our extensive knowledge of numerous biochemical reactions inside a cell and a complete lack of logical visualization regarding how all those reactions work together in an amazingly regulated order. To neutralize this gap, modern scientists have created a term that allows them to close their eyes on this gap and just ignore it. In a simplified example below, describing biochemical pathways in cells, I use this term in bold print – so you could see how scientific papers describe complex biochemical reactions.

On the surface of cellular membranes there are many thousands or rather millions of different receptors. These receptors are sitting on the membrane in such a way that usually one part of it is exposed to see the world outside, and another part is sitting under the membrane to transfer information into cellular "headquarter" (nucleus) to inform the genes about what is going on outside and how they should react to environmental changes. Suppose one cell has encountered a pathogen in the environment where pathogens are not allowed. Membrane receptor "A" has recognized a three-dimensional structure of that pathogen as a foreign intruder and sends **signals** to a protein "B" under the membrane. Protein "B" immediately reacts with protein "C" and the product of their reaction **signals** to protein "D," which turns into activated form, reacts with a dozen or more other molecules and sends **signals** to protein "E." Protein "E" is yet only a messenger boy in a long chain of numerous reactions. Each link in this chain sends **signals** to the next link and then to the next one and so on and so forth until finally the protein "Z" proudly presents the important message about the foreign intruder to those genes in the nucleus that are responsible for making important military decisions regarding how to organize the defense or how to commit suicide (apoptosis). Those genes immediately start **signaling** to numerous other molecules using different pathways and messengers. Finally, all these **signals** have resulted in an instant synthesis of specific antibodies or other "soldiers" that are released from the cell to find and destroy all unexpected invaders. Through many other **signals**, cells communicate with each other, tell each other how to behave, where to go, what to do. They teach each other, take care of each other and sometimes kill each other. All these things they do through **signaling** and biochemical reactions.

To give you a more accurate example of how modern scientific papers describe intercellular processes, I have selected one small excerpt from the latest issue of the European Journal of Cell Biology: "We have expressed proteins at various concentrations to analyze the strength of the **signal** that governs their subcellular localization. Our microscopic observations indicate that organellar sorting **signals** override the targeting preference of most cytosceletal proteins. Among these **signals**, the nuclear localization **signal** of SV-40 is strongest. The protein coronin, however, can only be misled by the nuclear localization **signal**. This correlates surprisingly well with the affinities of constituent **signals** derived from in-vitro experiments."

As I said above, we know a lot about those biochemical reactions, but if you ask any biologist how they visualize those signals, I guarantee that you will not get a satisfactory explanation. At best, they will tell you that signaling is a reaction between two or more molecules, which gives "green" lights to other reactions through their changed conformation or through a synthesis of new molecules with new properties. In my view, such explanations are missing a certain logic. Let's try to visualize these biochemical reactions from the viewpoint of simple logic.

A cell is not a sack of liquid fluid, where all molecules can flow freely in all directions by Brownian movement. A cell is a gel (like a jellyfish) and free movement of molecules just is not allowed in there, if the cell is alive. When the cell is dead, then small molecules can penetrate it by a simple diffusion, but for large proteins it is still not easy to diffuse into cells because even being dead, it is still a gel. Any biochemical reaction requires movement of its components. Protein molecules can interact with each other if they can move at least to the extent that would allow them to find those three-dimensional structures on their surface, called epitopes, which have sufficient complementarities for molecular forces to hold molecules together.

A receptor "A" on a membrane reacted with a pathogen and changed its configuration in order to give signals to protein "B." For successful signaling, at least two conditions are required. First, protein "B" must be always available in very close proximity to protein "A." Second, protein "B" must have enough flexibility for free movement in order to find that special epitope on protein "A." In a gel-like environment both conditions are restricted proportionally to gel density. This means that we just have to assume that for every imaginable biochemical reaction inside of a cell, all required components must always be available, ready and in close proximity to each other in order to provide efficient communication between the genes and the outside environment. Logically, it could be possible only if each cell works under very strict ORDER (or Directing Power, if you wish), which constantly oversees availability of all required molecules in all right places and positions. The only alternative way of thinking is free movement of all molecules, but it does not provide realistic picture of complex biochemical reactions because in gel-like environment the possibility of always having right molecules in right places is just against normal logic.

At this point materialists would say: "Of course, each cell works under strict Orders. This Order is controlled by all genes in the nucleus. They accept signals from all parts of the cell structure and react by synthesizing all required molecules for all required biochemical reactions." I just cannot confront such naïve arguments. Genes consist of the same matter as all other molecules in cells and they work under the same laws that all other molecules do. Unless we reconsider our view on molecules and atoms as inanimate matter, it is very naïve to believe that genes can be the only controlling and directing power of cellular functions. If there is nothing beyond matter and energy, then any cell would be just a sack of different molecules randomly reacting with each other by accidental contact. A live cell is not a sack of molecules, but a very complicated factory, and, like any other factory, it has a central headquarter, a board of directors, production facilities, offices, corridors, workers, messengers, managers, security, enforcement officers and even public relations and other highly specialized departments. All cellular elements are of inanimate matter (as we want to think) and nevertheless cells can move, breathe, multiply, talk to each other via chemicals, eat, defecate, show a certain preference for different conditions, contract on unfriendly substrates or relax in favorable environments. How can all these signs of life be attributed to only chemical reactions of inanimate molecules? This is just against common sense; even an atheist must admit that, if not God, then some kind of life force, which we do not know much about, must be responsible for the orderly conduct of numerous biochemical reactions in live cells.

To be convinced that modern biological science is missing something very fundamental and extremely important, you do not have to be an expert in biology or chemistry. All you need is an open mind, logic and common sense. Let me take you for a tour through biological facts that you already know from your school years, but most likely never asked yourself any questions about the nature of those facts. Let's go through the most incredible magic on this planet – the birth of a baby. We are all so much familiar with this process, that we do not see any magic in it. I will help you to look at it in different perspective and you will see (I hope) that the birth of a baby is a real miracle that absolutely cannot be explained merely by genes and chemical reactions.

A man produces about 300 million sperm every day. All of them are stored in the testicles; they are motionless in there. Each sperm looks like a tadpole with tiny head containing genetic codes and a long tail that can propel like a motor, making each sperm an excellent swimmer. At the moment when millions of them are released into the vagina, those motionless tadpoles suddenly wake up and become very energetic. Their tails start vigorously propelling and at this time all of them have begun an amazing race in one particular direction – to the egg. Have you ever asked yourself, at this point – How do they know what their final destination must be? How do they know what they have to do to achieve the task given to them? How do they know which direction to take to find the egg? If you asked your teacher these questions, the answer probably was: "These are all physicochemical reactions." Given this answer, have you asked yourself – How can chemical molecules that are inanimate matter be so smart to give those tadpoles lots of information in such a short period of time to get them to vigorously compete and to find their final destination – the egg in the uterus? This is only the beginning of the miracle.

All the sperm in vaginal environment express maximum energy and, pushing each other, they swim as fast as they can, trying to bypass all the obstacles of this races. The stakes are very high. The stake is Life for about 100 years in the beautiful paradise that is called the planet Earth. The lead swimmers are close to the finish line now. The egg is waiting impatiently. She did not waste her time waiting for the winner. She worked hard preparing her outer membrane in a special way to welcome the winner and to facilitate his entrance into internal chambers of a very sophisticated factory. Finally, the strongest, the fastest and the healthiest one hits the egg's membrane and binds its special receptor onto its surface. With the help of his tail, he quickly makes a hole in the specially prepared egg's membrane practically the same way as a woodpecker makes a hole in a tree with his beak. The egg is ready to embrace the lucky winner. Instantaneously after his penetration, she changes the structure of her membrane and closes all doors to make it impossible for other swimmers to make additional penetrations. The egg wants only one set of genes and as soon as she gets it, her assistants (other cells in the uterus and around the egg) clean the area to remove unlucky racing participants. At this point, Dear Reader, do you still think that all these very coordinated and strictly controlled events are just the result of physicochemical reactions controlled by genes? It does not sound like a logical explanation, does it? But the most exciting part of this miracle is still nine months in the future.

After the champion of the race successfully has penetrated the egg's internal structures, he is running to the egg's nucleus to deliver the long awaited package of genes in the form of Deoxyribonucleic Acid (DNA). However, the egg does not allow him to do it for a few minutes. "Look at yourself! – she says – "You just came in from a long run. You are all wet and dirty and… this smell… I know you love it, but the Board of Directors does not. You have to change before you go there with your business proposal". Saying that, the egg strips sperm of almost all his proteins and replace them with her own maternal proteins. All his paternal proteins she destroys with her enzymes and throws them out into uterus environment. "Now you look like a decent business partner! Go ahead, my Love, and start negotiating the nine months project and 100 years beyond!". With this blessing sperm fuses his membrane with egg's nucleus and releases his package of genes, being confident now that his partnership will be long and productive. Immediately after fusion with the nucleus, the tightly packed DNA molecules unwind into long strings in order to start a very complicated cycle of building their copies at every cell division, repeating it accurately trillions of times. They also have to start selecting which genes have to begin working now and which ones should wait for a later time. At the same time they start a certain competition or negotiation with matching egg's genes. If her genes make red hair, and his genes make black hair – which color hair will their child have? Who is deciding this – inanimate molecules of genes or some power that directs and controls their work? Do you think that all those extremely complicated decisions in the accurate coordination of zillions of biochemical reactions can be done again and again by molecules themselves without the involvement of any Directing Power? In this context the God Theory by Dr. Haisch (see Chapter "Open your mind") is definitely something to think about. But the magic is not over yet.

Up to this point, zillions of chemical reactions have already been performed in the egg in order to prepare it for a long nine month project. Zillions more biochemical reactions and trillions of different cells will be working in amazing coordination. Each new cell knows perfectly well what to do, where to go or how to help other cells to perform their duties. With this knowledge, a growing, formless conglomerate of cells starts to show the first signs of a tiny head, legs, hands and then fingers, eyes, ears. All these just because each cell somehow knows its role in a very complicated passage from a single cell to trillions of them, which at certain point come out in the form of a new baby to see the world and to enjoy the Life. This is the best and the most puzzling miracle God has created on this planet. Even, being raised as an atheist, would I ever believe that this miracle is only genes, molecules, matter and energy? When you think deeply about this and many other biological phenomena, you feel like you are enlightened and start seeing things in a different perspective, the perspective that convinces you that the real Intelligence is not merely an attribute of a human brain, but rather something outside of it. The human brain is just a piece of "hardware" in a biological computer, but somehow, somewhere, someone is controlling the keyboard.

Three "bricks" of the Universe

The Universe consists of billions of stars, galaxies and planets. They are dispersed in space at such distances that it is not easy to comprehend. The Hubble telescope has shown us the image of numerous galaxies that are 78 billion light years from our planet. And this is not the end point of the Universe. As we know, those endless galaxies are quite different, but to this context I would like to emphasize the facts of their similarity. We are always amazed by the endless variety of living organisms, minerals, planets, stars, galaxies, etc. However, we forget that all animals and plants, all mountains and rivers, all planets and stars, all galaxies and the entire Universe are built merely with THREE basic "bricks" - namely, protons, electrons and radiation. (I have omitted the later discoveries of even smaller particles since they are not relevant to our discussion. Besides, on the scale from big to small, there is no end on both sides). The first two "bricks" are material particles, having mass and size, while the third "brick" is presented by waves of radiation, which do not have mass or weight and are not material particles.

All three "bricks" can be transformed into each other, representing an integral unit in three different forms. This is an enigma that modern physicists, working with quantum mechanics, still cannot fully understand, but all agree that any particle of the matter can be a particle and a wave at the same time. Or in other words, any particle can be material and immaterial at the same time. Or yet in other words, any particle exists (material status) or does not exist (immaterial status) at the same time. To confuse you even more, I can say that every particle has anti-particle, and matter has anti-matter. This is not science fiction. Anti-matter is produced every day in two locations – the United States and Switzerland. The only problem is that it is being produced in such minute quantities, that it cannot yet be utilized by any technology. However, scientists believe that one day the use of anti-matter will allow human beings to travel on very long distances in space due to enormous amounts of energy that anti-matter can keep in a very small volume. I am not a physicist and it is very difficult for me to understand all these concepts, but if you are interested, you can find lots of information about it on the Internet. But, with this real science in mind, I would like to remind you now about two interesting analogies that are relevant to our attempt to build a bridge between science and religion.

Three faces of God

Christianity teaches us about three faces of God - God-father, God-son and God-Holy Spirit. The first two faces of God look something like material nature, like protons and electrons, but the third one looks immaterial like radiation (or a soul). This religion also insists that God is one in those three forms. It means that God can be presented by all three forms at the same time and can transform itself into each of these forms. At the time when such an image of God was formed in peoples' minds, nobody had any idea about protons, electrons or radiation. How did people come to such interesting images of God? Is it just a coincidence or did someone try to explain something when they did not yet understand physics?

Long before Christianity, people left us their image of the world sitting on three whales. Again this magic figure, THREE! I do not think that image of three whales supporting the Universe was taken by ancient people so literally. It was just a symbolic interpretation showing that in the essence of the Universe there are three major elements, three "bricks." Again, who tried to tell them, not having any knowledge about physics or chemistry, that the Universe consists of three elements? It would be interesting to speculate that this information was attempted to be transferred to people a long time ago, but because of the lack of knowledge it ended up in the form of three whales, or God in three faces? If we assume that some unknown Directing Power constantly transfers all kinds of information to people, then how can such information be transferred to people's minds and accepted in there? Let's try to answer this question by the method of exclusion.

Transfer of Information.

Since everything in this world consists of three "bricks," which of them can serve as a substrate for the transfer of information to Earth? The first two "bricks" are not qualified because being material particles they cannot be transferred very long distances without significant loss of energy. Besides, our atmosphere can efficiently detain these particles. If not electrons and protons, then only radiation is left.

Radiation is electro-magnetic waves that we can divide into five major categories in order of wavelength. 1) Radio waves (length 300 m - 60 cm), 2) Microwaves (length 60 cm - 0.6 mm), 3) Light waves including ultraviolet (length 0.0008 mm - 0.0001 mm), 4) Roentgen or X-rays (length 0.0000013 mm), 5) Gamma waves (length 0.00000000001mm).

All these waves penetrate the Universe in all directions like an ocean in which all galaxies, stars and planets are floating like tiny organisms. If the origin of life, its evolution and eventually the birth of an intellectual creature was happening under some instructions from a "Directing Power" coming to Earth from the Universe, then which of mentioned above waves could play a role as a transferring substrate? Consider the fact that all waves with lengths more than one millimeter are efficiently detained by the atmosphere, radio and microwaves are not qualified to be a transferring substrates, but they help us do a lot of things in our satellites and space ships. Then, if we take into consideration the earlier said assumption about at least the partial influence of a Directing Power in the work of our brain, we have to exclude light waves, since our brain can functioning in full darkness. Thus, Roentgen and Gamma rays are left. These high-energy waves can easily penetrate the atmosphere and Earth and can travel very long distances virtually without fading. Can these waves transfer information? Of course, they can, like any other waves.

We are comfortably sitting on a couch and holding the remote control, sending electro-magnetic waves to the TV set or VCR and making these machines do numerous operations, adjusting sound, color, focus, switching programs and so on and so forth. Virtually anything can be done now through the transfer of information a long distance by waves. No surprise? Of course not. We are too accustomed to this now. We send robots into space and to other planets and they do for us anything we want them to do via electro-magnetic waves sent by a computer operator. And we want to believe that in the entire Universe we are the only source of intellectual information. Isn't it time to reevaluate this snobbish attitude and try to look deeper at who we are? Our snobbism does not allow us to accept an idea that we might be just sophisticated machines controlled at least partially by a certain Power in the Universe. Let's take a look and see if our logic and science can support or destroy this idea.

Source, Carrier, Acceptor.

The transfer of any information from the Universe (or from a Directing Power) into our brain implies the presence of three elements: a) Source of information, b) Carrier of information and c) Acceptor of information. From these three elements, the nature of the Carrier of information does not cause any problems of comprehension. It can be represented by Gamma waves, or by something yet unknown, but of similar high energy. The acceptor of information is also easy to comprehend. It is probably some organic substrate that is located in the brain tissue of animals or human beings. Cosmic rays constantly penetrate our brain no matter where we are or what we are doing. Do they pass through our brain like water through a sieve or can they induce a certain excitation in some brain areas? Such excitations, as I wrote earlier, can induce some electro-magnetic signals that give us some images, activate some memories or make us perform some actions. This might happen in a complex interaction of radiation waves with two other "bricks" - protons and electrons in our brain. Such interaction ends up with certain signals going to our peripheral organs and making us behave in a certain way or transferring some information on a piece of paper or a blackboard in the form of recorded thought.

Let's assume for a moment that this is true. If it is so, then how we can imagine a transfer of information to Earth, when our planet did not yet have any living organisms capable of serving as acceptors of information? In other words, how did life begin? As all atheists believe, everything has been started from simple chemical reactions. Substrates for life might be developed gradually and there is no contradiction here with atheistic theories. After all, chemical reactions are just different interactions of two "bricks" - protons and electrons, while the third "brick" either comes out of the first two or is absorbed by them. Can electro-magnetic waves influence chemical reactions, induce them and create or participate in the creation of new compounds? Of course, they can! Every chemist knows that. But can they do their influence specifically? In other words, is it possible to develop complex compounds like proteins or DNA and later create live organisms by influencing reactions in certain directions with information coded in waves of radiation?

My knowledge of chemistry is limited, but I would like to hear what organic chemists would comment about such speculation. One way or another cosmic radiation could be quite active in physicochemical processes on Earth. When the first living organisms were developed, the mechanisms of Evolution with its main tool - Selection - started playing a more active role in further improvements and variations of living creatures. The role of evolution and selection cannot be underestimated. The real arguments start when we begin to discuss the accidental nature of events or directed ones.

Thus, an Acceptor of information can be possibly presented not only by such sophisticated organs as the human brain, but also by simple chemical compounds, which also consist of the same three "bricks" – protons, electrons and radiation. With a certain amount of fantasy and logic, we can imagine now how information, coded in waves of radiation, could be accepted by molecules and could lead to some chemical reactions that in their turn step by step could create more complex forms of matter and ultimately a primary living cell. But how can we imagine the third element in the transfer of information - namely, its Source? This is the most difficult area for any speculation since we do not have any scientific foundation that could support at least partially any fantasy in this field. Our only chance here is to use our logic and to keep our minds open. Let's try!

What do we know about the Sun?

What is the Sun? Where did it come from? How come it has radiated light for five billion years and will continue doing so for another five billion? Does it provide us just warmth and light or something else that we are not capable yet to comprehend? Let's see what scientists can tell us about it. They say that the Sun is 109 times bigger than Earth and consists of hydrogen and helium. Moreover, helium occupies the central core of the Sun where it splits and becomes hydrogen releasing an enormous amount of energy. There are billions of stars like the Sun in the Universe. They are like live organisms; they are born, live and die. Their "life" depends on substrates of potential energy (probably helium, like in our Sun), but they can take energy from surrounding planets pulling and swallowing them. Scientists say that this destiny is inevitable for our planet as well. When it happens, Mars will become closer to the Sun and frozen water will melt and will create favorable conditions for life development. Then with the help of Evolution and a Directing Power, new life will create another intellectual creature on Mars. Maybe even before a man will be able to settle on Mars and start its destruction, like he has done on Earth. It would give him a chance to extend his presence in the Universe for a few more billion years until Mars becomes the next victim and is swallowed by the Sun. (Most likely, however, our civilization will destroy itself being unable to put under control overpopulation of this planet.)

Open your mind.

We believe in the existence of intellectual creatures living somewhere on other planets of the Universe. This belief does not have any scientific foundation besides logic. But somehow this belief has a certain degree of conservatism, which does not allow us to imagine an intellect on the other planets outside of the brain or even in another form of matter without protein, DNA or organic substances. If human intellect is basically the work of three "bricks" – electrons, protons and radiation, then why can we not imagine a similar work in a different form of matter – let's say in the form of the Sun. The ravings of a madman? Maybe. **However, any science would be stagnant, if someone sometimes somewhere would not ask somehow some "crazy" questions!!!**

Nobody would argue that the Sun is responsible for making conditions on Earth suitable for the development of life. But we know so little about the Sun and how it works that any information about it is highly speculative. Scientists say that thermonuclear explosions of enormous magnitudes are going on constantly on the Sun and inside of it. If this is so, why does the Sun not explode at once like a huge hydrogen bomb? Every second it turns into energy only five million tons of its mass doing it accurately for five billion years. With such rationality it seems that all explosions on the Sun are not so chaotic. If they are not chaotic then a certain kind of order possibly maintained by a Directing Power has to be assumed.

The sun consists almost entirely of three "bricks" in there purity. The same "bricks" from which we are built and from which everything around us is built. Helium (which is a proton and two electrons) is located in the center of the Sun, and hydrogen (which is a proton and one electron) is on its periphery, and virtually there is nothing else. Helium is split by thermonuclear reactions and becomes hydrogen plus radiation. All these reactions give our planet light, warm temperature

41

and a lot of radiation. This constant transformation of material substance (the Sun's mass) into immaterial (radiation) will last another five billion years. Such incredible efficiency of the use of the Sun's mass is hard to comprehend as just a chaotic process, but rather must be a highly ordered chain of reactions controlled by some Power we know nothing about, yet.

If the Sun represents a highly ordered structure of events resulting in a constant release of radiation directed to our planet, then wouldn't it be crazy to assume that Sun actually IS that Directing Power, the Directing Power that I am talking about through this whole book? If so, then maybe the Sun's radiation carries a lot of information influencing not only the general development of life, but also behavior and even thoughts of all inhabitants of our planet.

Fantasy? Of course, it is a fantasy! But give me one reason that can show me that this fantasy is just an absurd one. One could argue, pointing to the fact that each piece of light (photon) requires eight minutes to cover the distance from the Sun to Earth and several million years to cover the distance from other galaxies. At first glance, it would create a certain inconvenience for those speculations that assume that some information is encoded in photons and delivered to the brain for any particular action. However, according to Einstein's Theory of Special Relativity, it is not quite true. This theory claims that each photon, traveling with the speed of light, is instantaneously absorbed by my eye or any other object on its way regardless of any distance between its origin and a final destination. Dr. Haisch, whose book I will be talking about below, says: "That's because, in the reference frame of a particle traveling at the speed of light, all distances shrink to zero and all time collapses to nothing. From its own perspective, the photon of light leaps instantaneously from there to here because distance has no place in its existence." Well, I am not a physicist and for me it is not easy to imagine the disappearance of time and space, but I believe in Einstein's logical thinking and in his theory. Thus, the argument mentioned above is now disappearing with the speed of light.

If we accept the idea of the Directing Power controlling all events on our planet, what other than the Sun's sources of radiation could efficiently send information to Earth by means of radiation waves? In certain degrees other stars also can send radiation to Earth. (In this respect predictions of astrologists might have some sense in those cases, when they are not made just for our entertainment). It is not easy to imagine a "fire ball" (our Sun), which is 109 times larger than our planet, but other "fire balls" in the Universe are thousands of times larger than the Sun and they release a lot of radiation in all the directions of the Universe. However, our Sun is the main source of radiation in our neck of the wood and its influence on Earth is much greater than from other stars.

Thus, if we accept a hypothetical role of the Directing Power in the influence or maybe even direct control of our activities on Earth, the mentioned above three elements (Source, Carrier, Acceptor) needed for the transfer of information are presented by the Sun, Radiation and the Substrates of our Brain. In this context it would be relevant to remind you that ancient Egyptians did not have any idea about radiation, but they believed firmly that the Sun was their God and worshipped it with veneration.

The thought described above is not an attempt to build a hypothesis or theory. It is just one, maybe crazy, thought in an attempt to find answers to numerous questions about life and the human mind.

Recently one book with a very striking title "The God Theory" attracted my attention. There are an endless number of theories about God and most of them are written by religious fanatics. However, this one was written by Dr. Bernard Haisch - a scientist astrophysicist with a very impressive background. Usually, when highly educated and experienced scientists write their views about God, life, intellect or the Universe, I cannot resist my temptation to read their viewpoints in hopes to find a rational explanation of all these miracles that surround us and of which we are a part. So, I bought the book and read it cover to cover.

I was not disappointed despite the fact that I have not found the answers to my questions, but I actually did not expect to find them. The book has quite enough historical and scientific information that certainly was interesting to read, but I was looking for the God Theory. Finally, I have found it. Here it is:

"In the God Theory, consciousness is the primary stuff of reality. Consciousness is able to shape and direct matter. Consciousness, in fact, has created this universe – the planets and stars, the plants and animals, and you and me. The initiating consciousness creates your whole world for its

43

own evolution, its own growth, and perhaps, its own amusement. This is the essence of the God Theory.

The physical universe and the beings that inhabit it are the conscious creation of a God, whose purpose is to experience his own magnificence in the living consciousness of his creation. God actualizes his infinite potential through our experience; God lives in the physical universe through us. Our experience is his experience because ultimately we are him, that is, immortal spiritual beings, offspring of God, temporarily living in the realm of matter."

If this would be written by Deepak Chopra or a priest, I would not be surprised at all, but that was the viewpoint of a high caliber scientist and I was perplexed. When my initial shock was gone and my thoughts about his ideas were still boiling in my mind, I gradually came to the conclusion that Dr. Haisch actually gave us a quite interesting speculation, but he was afraid to express it in a clear bold form. If I am right, I perfectly understand his fear, since his career still depends very much on his relationship with the conservative scientific establishment that does not allow him to deviate from the fundamentalist reductionism denying everything beyond matter and energy.

Still, trying to keep my mind open, I have decided to take his ideas and to express them in rather bold speculations. I am a retired scientist and do not care if fundamentalists throw my book into a trash can. Dr. Haisch may or may not disagree with my interpretation of his ideas, but I still believe that this is actually not mine, but his interpretation, that he was afraid to express clearly due to the complete absence of supporting facts. Another reason I want to write about it is because I think that at least partially his ideas could be tested in a laboratory environment and I will talk about that a little later.

The brainchild of his book is not actually the God Theory, but the zero-point field inertia hypothesis, which is the core of his ideas. In my attempts to interpret his God Theory, I have to explain first the zero-point field and I will do it by quotations from his book:

This "is essentially the story of light – a very special light known as the electromagnetic zero-point field, or the electromagnetic quantum vacuum." "Radio, television and cellular phones all operate by transmitting or receiving electromagnetic waves. Visible light operates in the same way, just at a higher frequency. At even higher frequencies, beyond the visible spectrum are ultraviolet light, x-rays, and gamma rays. All are electromagnetic waves that are really just different frequencies of light." "According to Heisenberg Uncertainty Principle, at every possible frequency, there will always be a tiny bit of electromagnetic jiggling going on. And if you add up all these ceaseless fluctuations, you get a background sea of light whose total energy is enormous. This is the electromagnetic zero-point field." "Zero-point refers to the fact that, even though the extent of this energy is huge, it is the lowest possible energy state. All other energy operates over and above the zero-point state. Take any volume of space and take away everything else – in other words, create a vacuum – and what you are left with is the zero-point field full of zero-point energy. We can imagine a true vacuum, devoid of everything, but in the real world, a quantum vacuum is permeated by the zero-point field with its ceaseless electromagnetic waves." "The world of light that we can see is all the rest of the light that exists over and above the zero-point field." Thus, "the Heisenberg Principle mandates that all of space must be filled with zero-point energy."

As I understand this concept, we are like fish, we live surrounded by water, but our water is zero-point energy. "It acts as a kind of blinding light everywhere, inside and outside of us, permeating every atom in our bodies."

Well, at this point this concept is already quite interesting, since it shows us that all our cells, all our molecules, all our atoms, electrons, positrons, neutrons, quarks, etc., are submerged in a "water" of energy. The natural question comes up from this concept – How does this energy affect or influence the extremely complicated work of all those molecules and atoms in our cells and especially in our brain cells? Is this energy just an inert environment for us, like water for fish, or can it do much more that we still cannot appreciate and estimate? Dr. Haisch gives us a partial clue to these questions:

"Einstein's special relativity theory tells us that light propagation defines the properties of space and time. I argue that light propagation may actually create space and time. The zero-point field inertia hypothesis implies that the most fundamental property of matter, namely mass, is also created by light."

This hypothesis, described to date in several prestigious scientific journals and supported be some other high caliber physicists, does not look to me like just another crazy speculation. Thus, if light can create mass, then the Big Bang and numerous other scientific theories start making more sense and give us ground for other interesting ideas. This, I believe, was the main purpose for Dr. Haisch to write his book – to describe his hypothesis rather than the God Theory. After all, if he would title his book "Zero-point field inertia," I would not buy it and the majority of other readers would not do so as well. Incorporating his hypothesis into the God Theory was a very smart move, but by talking about God, consciousness and zero-point energy, he is afraid to speculate that the Intelligence that many people would like to find outside of our own brains, could be represented by that zero-point energy, which you can call God, Consciousness, Light, Love (oh, no, not Love, please!), or as I call it in this book, a Directing Power.

The work of a single cell (see Chapter "How do cells work?") or their assembly of many trillions of cells in the human body is an extremely coordinated process. This process created the Mona Lisa and sent spacecrafts to distant planets. But believing that this process is only the result of biochemical reactions of zillions of atoms and molecules, seems to me so naïve, that I just cannot believe that this concept prevails in the minds of many talented scientists. They still deny any necessity for a certain Intelligence outside of the human brain and still believe that there is nothing beyond matter and energy. Dr. Haisch believes in such Intelligence, but he has chosen to talk about it from a rather religious viewpoint, calling it "Divine Consciousness," "His Magnificence", etc. If he would try to make a direct connection between His Magnificence and Zero-point energy, then his God Theory would make better sense. On the other hand, by doing this he would probably close the door to government grants for his more scientific and less divine work.

To correct his oversight, I would like to see his zero-point energy as a possible source or a substrate of the Intelligence. We certainly know absolutely nothing if there is any ground to support this idea, but Dr. Haisch gave me one preliminary idea for some laboratory tests. Later in the Chapter "What is Intellect? I expressed my desire to build a chamber that would not allow any electromagnetic waves to penetrate, including zero-point energy. Dr. Haisch wrote (if I got it right) that a certain setting between two metal plates can remove zero-point energy in the space between plates, but that space can be only a few millimeters. Such a narrow space might not be enough to study even small animals, but could be sufficient to study cell behavior in a culture medium. Naturally, the interpretation of any results from such experiments most likely would not give direct support to the God Theory, but at least some influence of zero-point energy on cell behavior could be either confirmed or put aside as a force with no effect. For me the God Theory was interesting from the viewpoint of identifying God with electromagnetic energy that can create mass and be material matter and immaterial at any particular moment of time.

Soul, how real is it?

Believing in Soul is so natural that the image of it was probably developed long before any religion, in the era of the most primitive people. If you ever witnessed a sudden death, your brain would probably have the same old question: "How come! – Just a minute ago this still body was running around, making jokes, laughing, and gesticulating, and now it is completely motionless and never will do those things again?" Comparing these two stages of the same body, you clearly can see that a minute ago this body had "something" that was giving it the ability to perform numerous tasks, and suddenly this "something" has gone, leaving this fast moving body like a stuffed manikin. All his organs are still alive and all his cells still continue doing their usual biochemical reactions. In a short period after death, each organ, including the heart, can start functioning again, if transplanted into a live person. Each cell of any tissue can be taken from the dead body shortly after death and continue to live in a special medium, showing the full spectrum of live functions. It looks like all parts of this a minute ago energetic machine are still alive at the moment of death, but something invisible is missing and all these parts cannot move without that illusive detail, which has suddenly disappeared. Such natural thoughts have created an image, which we now call the Soul.

Nobody knows for sure what the Soul is all about, but definitely it is some kind of life force, which distinguishes live organisms from dead ones. In a similar way, any mechanical computerized robot becomes just a bunch of metal hardware as soon as one takes out its battery – the source of energy. Extrapolating this example to biological robots – human beings – we can assume that a Soul is a source of energy, which is responsible for all functions of live organisms, but leaves the body when the heart stops pumping blood.

The history of human civilization has an endless number of stories, which are supposed to prove the existence of a Soul. However, even keeping my mind open, I do not see much credibility in the overwhelming majority of these stories. The recently published book "Soul Proof" by Dr. Mark Pitstick presents a big collection of stories about the Soul, Reincarnation and Life-After-Death. Naturally, I did not find any credible proof of the Soul's existence, but his collection of stories is quite interesting to read for religious people as well as for agnostics. Dr. Pitstick himself is a very religious person and I feel that he genuinely believes in all his findings. The majority of them fall into a category of anecdotes like "swear to God" stories about the spirit of a deceased family member informing survivors about where they can find a hidden insurance policy. But I was looking for those "proofs" that could show at least a bit of science.

Among his stories, the information about reincarnated souls, collected by psychiatrists from patients under hypnosis, has attracted my attention. Hypnosis is certainly a science despite the fact that nobody can explain it since nobody knows what the human mind is all about and especially how it works. However, from modern psychiatry and biology we know that the human brain consists of two hemispheres and while the left brain (the conscious mind) is analytical and responsible for virtually all our day time activities, the right brain keeps the subconscious mind and works constantly day and night keeping all our physiological functions in the right order to ensure that we remain alive and healthy. The most intriguing fact of psychiatry is the ability of a hypnotist to turn off and on the conscious mind (the left brain) and to communicate with the subconscious mind of a hypnotized patient. Hypnosis is a well known scientific fact, but it is still so bizarre for all people that this phenomenon creates a lot of mysterious stories and a lot of groundless speculations. Nevertheless, this phenomenon might be the most rewarding in the search for the answers about the nature of the human mind or the existence of a soul. For this reason, I was trying to find at least a few "pieces of gold" among the huge collection of information from the experience of many hypnotherapists.

I picked up the book written by Dr. Brian L. Weiss, M.D. *"Many Lives, Many Masters"* and read it with great interest to the last page. Dr. Weiss has long-time experience in the field of hypnotherapy and his general background seemed to be quite impressive. When I finished his book, I could not and still cannot figure out if his book is just a fiction to attract more patients to his private practice, or a real gold mine of information, with an absolutely priceless value. There is no doubt that Dr. Weiss is a good writer, but to take the experience, described in this book without doubts, would not be easy.

In his book, Dr. Weiss introduces the reader to one unusual patient – a young woman who was looking for help to cope with her psychological problems including panic attacks and extreme anxiety. Trying to find the source of her symptoms, Dr. Weiss applied the so called "Regression Hypnotherapy." This therapy allows the doctor to find, in the subconscious brain memory of the hypnotized patient, those life events that might be responsible for their psychological problems, and then remove them. This whole book describes his experience with only one woman named Catherine. This experience was so unusual that according to Dr. Weiss' own words, it turned his own life upside down. For eighteen months Catherine had intensive psychotherapy, coming to Dr. Weiss once or twice a week for hypnotic sessions. During those sessions she revealed, under hypnosis, her experiences from past lives, when she lived being sometimes a woman and other times a man. In a series of trance states, she also acted as a conduit for information from some kind of "spirit entities," which are called "Masters" in this book. Here is a short statement from Dr. Weiss about his experience with Catherine:

"Nothing in my background had prepared me for this. I was absolutely amazed when these events unfolded. I do not have a scientific explanation for what happened. There is far too much about the human mind that is beyond our comprehension. Perhaps, under hypnosis, Catherine was able to focus in on the part of her subconscious mind that stored actual past-life memories, or perhaps she had tapped into what the psychoanalyst Carl Jung termed the collective unconscious, the energy source that surrounds us and contains the memories of the entire human race.

This book is my small contribution to the ongoing research in the field of parapsychology, especially the branch dealing with our experiences before birth and after death. Every word that you will be reading is true. I have added nothing, and I have deleted only those parts that were repetitions."

At the end of his book, Dr. Weiss added: *"I am no longer concerned with the effect this book may have on my career. The information that I have shared is far more important and, if heeded, will be far more beneficial to the world than anything I can do on an individual basis in my office."*

These remarks first seemed to me like coming from his heart. Considering myself as being already too far from blind atheism, but still too far from any religion, I wanted to believe Dr. Weiss and his interpretation of Catherine's revelations. However, when I read from his other book about his close friendship with Jesus Christ during one of his past lives and a miraculous cure of cancer patient (allegedly as a result of that friendship) by applying hypnotherapy, I got a better understanding what I am reading.

Nevertheless, the outright dismissal of something we currently cannot explain, does not help us in the search for truth and I decided to continue my search with an open mind.

I picked up several other writings by prominent psychiatrists-hypnotists and read their experiences. All of them have huge collections of similar data in support of past lives and reincarnation. Some of them are supporters not only of past life regressions, but also future life progressions, which are much more difficult to accept since there is no way to verify the progression data. Regression data sometimes can be verified and there are many stories in literature that allegedly have been verified and confirmed. Believing in those stories is a personal choice. However, one category of those stories could be an irrevocable proof of reincarnation and immortality of a soul. Those include the cases, when children, who have never been exposed to foreign language, reveal the knowledge of different languages under hypnosis depending on which past life the hypnotist could bring their subconscious memories under in a deep trance. This phenomenon is called Xenoglossy. In my personal opinion, this kind of evidence would be as sensational as the landing of a UFO with live aliens on the lawn of the White House. Then, why can't I find a single videotape of such children, but only stories from their psychiatrists? I certainly realize the potential danger of such videotapes, but they could be prepared with full concealment of the children's identities and still in the presence of reputable witnesses. If any reader of this book is aware of such videotapes, please, let me know by email at intart@peoplepc.com

Another very interesting set of evidences about soul existence comes from two well known phenomena – Near-Death-Experience (NDE) and Out-of-Body-Experience (OBE). Both of these experiences have been claimed by many people worldwide, but verification of those claims is usually very difficult or often impossible. There are many allegedly verified stories in literature, but again, believing those stories is just a personal choice.

The NDE stories come from people who have been dead for a few minutes, but then have been returned to life by the efforts of physicians and modern technology. This period of time usually lasts from 5 to 20 minutes, and after that the brain becomes irreversibly damaged. These few minutes are usually enough for your soul to look from above at your dead body and the physicians around you, to make a quick trip to the Other World, to say "Hello" to Jesus or Mohamed, to hug your dead relatives or friends, to see what you can expect when you are irreversibly dead, and to return back into your body with exciting memories, if physicians succeed in their efforts to bring you back to life. These stories would not attract much attention from serious scientists if they a) would not be too numerous to dismiss outright, b) would not have too many surprisingly common features, like narrow dark tunnels with a very bright light at the end, guides, relatives, similar landscapes, etc. c) would not be possible to verify by some facts that could not be seen by the dead body at the moment of death, but could be "seen" by the soul that just left the body.

A similar ability to see your own body from above or to leave your body for a while and go to your friend's house or other places, was claimed by many people, who have not died, but had OBE caused by circumstances or by meditation. If I had not had a chance to read one quite scientific book (see below) about NDE and OBE, I would be still very skeptical about such claims.

We should not forget about some natural features of a human character. I remember how one girl claimed on a TV show that she was pregnant by an alien from a spaceship. If I would have enough money, I could supply every TV show with hundreds of girls, who would be willing to describe in all embarrassing details how they got pregnant by aliens. For the same reason the majority of NDE and OBE claims are dismissed by the scientists, who conduct investigations of these phenomena. However, one particular category of these claims cannot be easily dismissed. This category includes stories described by people who are congenitally blind, but claim clear visual perception under NDE or OBE. Imagine for a moment that someone never had a chance to see light, being born completely blind. If such a person experienced NDE or OBE and could describe such an experience in all the same details as a normal person with normal visual perception, then the natural question comes to mind: "How is it possible?" In any library you can find books with many stories describing the "facts" that the soul of a blind person is not blind. The soul can see and if it leaves the body of its blind host for a few moments and returns back, it can bring memories of what it could see around. Is this realistic?

Two scientists decided to investigate this phenomenon using interviews, verification, when possible, and statistical analysis. This is the only serious scientific study I could find so far about NDE and OBE. The second edition of their book was published just recently in 2008. The title is: "Mindsight. Near-Death and Out-of-Body experiences in the blind" by Kenneth Ring and Sharon Cooper. I recommend this book to anyone who is seriously interested in this subject. In the first half of their book, the authors present impressive collection of data and interviews, and in the second half they analyze data with critical but still open minds. Briefly, their research shows that claims of blind people about visual perception under NDE and OBE are statistically correct and remarkably similar. However, in their interpretation of this phenomenon, they cautiously try to separate the definition of visual perception for sighted people from that of blind people. Despite some of my disagreements with their conclusions, I was glad to see that there are still scientists in some places, who do not reject those ideas that are not in line with official views of the scientific establishment and do not have any chance for financial support except private contributions.

My own analysis of NDE, OBE and hypnotherapy data only have convinced me again, that there is something very fundamental beyond matter and energy, and it absolutely cannot be limited merely to biochemical reactions in the brain.

The logic tells me that some kind of energy leaves the body when the heart stops. As I mentioned above, from the moment of death and at least to the moment when the body temperature drops too far down, every single cell in the body, including heart cells and brain cells are still alive and functioning, but the body is not. This particular fact makes me to believe that somehow the "battery" supplying necessary energy is gone from the body and as a result of this the brain stops sending signals to the heart or the heart stops reacting on those signals. What is this energy? Where has it gone? Why cannot modern technology identify it as an independent package of quantums, photons, or whatever? If, according to Dr. Weiss, a Soul keeps past memories, it must be an integral package of energy that can keep its personality, its characteristics of past lives and it cannot blend to the zero-point energy surrounding all of us in the Universe. In the context of the previous chapter about the Sun and some speculations about it, I would like to quote one of Catherine's revelations under hypnosis by Dr. Weiss, when she was dying in one of her previous lives: *"…For now I just feel the peace. It is a time of comfort. The soul… the soul finds peace here. You leave all the bodily pains behind you. Your soul is peaceful and serene. It is a wonderful feeling… wonderful, like the sun is always shining on you. The light is so brilliant! Everything comes from the light! Energy comes from this light. Our soul immediately goes there. It is almost like a magnetic force that we are attracted to. It is wonderful. It is like a power source. It knows how to heal."* Isn't she talking actually about the sun as a "power source" that attracts all souls by "magnetic force?" Everything comes from it and everything goes to it. Her revelations are quite in line with my speculations. Who knows? Maybe there is a grain of truth in this and maybe one day scientists will be able to tell me after my reincarnation "Who created the Mona Lisa?"

Since there are no satisfactory answers to any questions about soul, the dream of religious people is going, without any restrictions, in all directions. People want to believe in the stories about Life-after-Death, because nobody wants to die and maybe never experience again that divine gift of God that we call LIFE. I am still very skeptical about Life-after-Death. Even if there is some kind of existence after death, it is not life as we perceive it when we are still alive. In my definition, life is a complexity of your conscious memories of your experiences from birth till death. Your kindergarten years, your school years, your relationship with your parents, your friends, your first kiss, your first love, your spouse, your kids, your interesting trips or vacations – all these things is your life experiences. Take any old man, remove all memories from his brain and ask him: "Did you have good life?" The only response you could expect from him would be: "What is life?"

People's fantasies about life-after-death are so remote that sometimes they takes the form of laughable anecdotes. Muslims, for example, believe that for killing "infidels," they will be rewarded in heaven with forty virgin girls. This is a natural dream of young men whose culture puts so many restrictions on their sexual behavior and treats women as just sex objects. Christian's fantasies in this field are not less laughable. They usually have enough sex, but not enough money and therefore their description of life in heaven concentrates mostly on lots of gold and precious stones in and around the palace of God. Here is an excerpt from Revelation 21:11-21: *"The wall was made of jasper, and the city of pure gold, as pure as glass. The foundations of the city walls were decorated with every kind of precious stone. The first foundation was jasper, the second sapphire, the third chalcedony, the fourth emerald, the fifth sardonix, the sixth carnelian, the seventh chrysolite, the eighth beryl, the ninth topaz, the tenth chrysoprase, the eleventh jacinth, and the twelfth amethyst. The twelve gates were twelve pearls, each gate made of a single pearl. The great street of the city was of pure gold, like transparent glass."* This city is only 1,500 miles high in the air. I think that tourist agencies should pay more attention to organized tours to this wonder-city using modern space technology.

Whoever wrote this does not even understand the simple fact that if all roads in heaven are covered with gold, then it must have the same value as dirt on this planet. The Bible is the richest collection of people's fantasies. I am not going to criticize this book here, since it is so important for so many people who believe that the Bible was written by God. In the context of my book I could even agree with that if human brain is indeed just an acceptor of signals from outside. In such a context, the Bible was written by God and so was Hitler's "Mein Kumf," and so were many atheistic books and so is my book, and so are all those books on Amazon.com. It looks like God is a very prolific writer with very diversified and controversial ideas. Why is God doing this? I do not have an answer to this question, but if you believe in God, I can only say that "God's ways are unpredictable."

Is it possible to assume that God has nothing to do with all books ever written on this planet and all of them are written by human beings of their own free will? If we assume that, then we have to start believing that all molecules and atoms in the brain are so smart that by immediate chemical reactions they send zillions of electrical signals to the muscles of my hand that moves my pen in a very strict manner resulting in words and sentences written on this piece of paper. Is that a more realistic assumption? Not to me! Is there any other way to explain who is the real author of all books ever written? At this point I do not see any other way.

The image of a Soul as an independent package of energy is strongly supported by Asian culture. Chinese and some other Asian people firmly believe that each organism has a certain package of energy sitting locally inside of an organism. This energy is called "Chi" and it can be moved to other parts of an organism under certain conditions. This idea of a local package of energy is used by modern Chinese medicine and is not denied now by some not so skeptical Western medical professionals. "Chi" is responsible for all physiological processes in an organism. Those, who managed to control "Chi" successfully, use it for correcting medical disorders. Assuming that our brain is just a substrate-acceptor for immaterial energy carrying information, I was speculating above that all our actions might be determined by that information and coded in electro-magnetic waves.

However, an alternative speculation may assume that the whole package of such immaterial information is stored in material substrates of an organism in the form of energy, which the Chinese call "Chi," and we call "Soul." If so, what happens with that energy (or Soul) after death? Death is actually a disintegration of the outdated material substrate. If the Soul is not material, it cannot die in the same pattern as the material substrate. Indian philosophy says that the Soul is transferred into another substrate and the new substrate can represent an organism of a different species. For example, the human Soul can be transferred into a plant or an animal. For such a conception we have to assume that the Soul is an integral form of energy like a photon or quantum, and it cannot be dispersed in space like material particles (e.g. air). If so, we have to admit the wisdom of all religions that always point to immortality of the Soul. If the soul, according to Dr. Weiss, can be traced thousands of years back through regression hypnotherapy, it certainly must be in some form of integral energy package. Looking at the soul from this perspective, I recall another high quality videotape made by journalists from "60 Minutes." This tape shows two small boys. They are brothers from the same parents and have a very small age difference. Each boy has his separate room, separate toys, separate books, separate games. One of them behaves like a man. His toys are solders, tanks, airplanes, guns, etc. If you enter his room, you can see right away that this is a boy's room. If you enter the room of his brother, you unmistakably can see that this is... the girl's room. His toys are Barbies, dolls, soft animals. All the decorations of his room and colors are a typical girl's decor. And when he talked to the journalist, he was shy and soft like a typical little girl. Both brothers were growing up in the same environment and had equal treatment. How can anyone give a reasonable explanation to their differences? In the context of this book, it would be quite logical to speculate that the female soul somehow (maybe by mistake) entered the body of one of these brothers. Could this also explain the certain features of gays? I do not know, but I am surprised that our modern technological progress still cannot identify the soul as an integral form of energy controlling all the processes in organic substrates and by doing this giving it life. Could this be because the three basic "bricks" – protons, electrons and radiation represent an integral unit in three forms constantly converting into each other and being unable to exist without each other? If we were better able to trace the transformation of these three elements of the Universe, we could possibly better understand many of the puzzles of our lives.

Passive Acceptors or Reciprocal Communicators?

The following section of this book was written much later, namely, on February 4, 1996; that is about seven years after the main part of it was finished. I was too busy doing science at the National Institutes of Health in Bethesda and did not have time to think more about the puzzles described in this book. However, on the night of February 3, I suddenly woke up (the clock showed 3:35 am) because of one thought that struck my mind. The thought was so powerful that I could not fall asleep again. It supported my previous discussion about a Directing Power, giving me some new interesting analogies between live organisms and the Sun. It also supports an idea that our brain is not just a passive acceptor of information, but a machine having constant exchange of information with a source, which I call the Directing Power.

We are quite accustomed to seeing a halo around the heads of saints in icons and religious paintings. Until recently we thought that this was just the imagination of religious people. Now we know that the halo is real and it belongs to any live organism, but we call it "aura." Aura is nothing more than radiation, going out of our body and especially out of the brain. It can be registered by modern photography and it can be even seen visually by a very small percentage of people. At the present time, aura is being studied seriously and some attempts have been made to use aura in diagnostic procedures.

Here I want to remind you again that the biochemical composition of the brain is poor in comparison to many other tissues. However, despite the simplicity of its content it can momentarily accept and process an enormous volume of information. In the examples of blind musicians or man calculators described above, light and sound waves induce immediate recording of information, its digestion and transfer to peripheral organs for a certain performance or writing or whatever signals that digestion can generate. Moreover, this immediate processing of information does not require any conscious attempts of thinking (during computation) or memorizing (during recording of a concerto). Thus, the brain does this incredible work automatically without any WISH or THOUGHT or any attempt from our consciousness.

Logically, I was suggesting above that a certain Power is supposed to direct this "automatic" work. For a religious person this Power is God, and for a scientist-materialist this Power is a set of personal genes in the brain of an individual. In other words, the arguments are still in the same area, whether this Power is only inside the brain or, on the contrary - outside of it. If it is only inside a brain, then we have to assume that processing any information is regulated somehow by a complex chain of biochemical reactions in the brain. Taking into account a relatively biochemical simplicity of brain tissue, it is virtually impossible to imagine that whole piano concertos can be recorded, processed and reproduced with unlimited improvisations by a chain of biochemical reactions inside of our neurons. Believing that biochemical reactions can be so smart to create the Mona Lisa or Mozart's concertos is not smart at all. Most likely biochemical reactions do nothing at all in such sophisticated digestion of information, just as they do nothing in the processing of information by an electronic computer, which can do much of the same work without any biochemical reactions. Biochemistry of our brain is probably needed merely to maintain the brain in its full ability to accept signals, as a computer's hard-drive has to be maintained in good shape to accept signals. But again, the computer takes signals from our hand; and which hand sends the signals to our brain?

According to speculations in this book, our brain is built in such a way that three major "bricks" of the Universe take their mutual transformations in the brain all the time constantly releasing radiation into the Universe and accepting it from the Universe. This radiation, going at the speed of light, could possibly represent those strings that connect the Directing Power to all live computers on our planet.

Thus, an Intellect might be not a peculiar quality of a human being, but a property of the Universe that works somehow through mutual transformations of the three major elements – protons, electrons and radiation. Looking at photographs of human auras, the following analogy comes to mind: Mutual transformations of protons, electrons and radiation happen all the time in our head and the same process is going on constantly in the Sun. Bright and colorful auras can be seen around human heads and the same aura can be seen around the Sun. Thus, every human head is like a tiny imitation of the Sun created by it, maintained by it and possibly controlled by it.

Enlightenment, Vision and Communications with Spirits.

The above speculations can shed light onto many unexplained phenomenon of human psychology and build bridges between science and religion. When I woke up in the middle of the night by the thought about the analogy of the brain and the Sun, I was perplexed - "Why did this thought suddenly come to my mind, when I did not think about it for years? Who sent this information into my brain? Is there any sense in such unexpected awakeners or it is just an accidental reaction of some molecules in my brain?" Maybe this is the same way the information about the three faces of God was transferred to people's minds when nobody yet knew about the three "bricks" of the Universe. We do not have answers to all these questions. However, many scientists have personal experiences with phenomenon of "enlightenment." They may work for a long time trying to resolve a task or get a better understanding of something unusual. Quite often the final understanding or a key decision comes at nighttime, when the brain is in deep sleep. How is "enlightenment" elaborated in the brain? Are these accidental biochemical reactions or deliberately transferred information?

Atheists do not deny the phenomenon of enlightenment (they experience it too), but they explain it on the basis of their religion. "Your brain – they say – continues working, when you are sleeping. Those major tasks that occupied your brain during the day are still in the brain's memory and the brain continues digesting them under more favorable conditions, since at night there are many fewer destructive forces that take brain energy than during the day."

Isn't it wonderful?! I am in sound sleep, snoring for the entire neighborhood, but my brain, my faithful old chap, still works for me, trying to resolve my daytime puzzles. If this is so, I want to know how all my brain molecules can work so hard without my involvement, when I am as useful in my bed as a piece of wood. How do all those protons, electrons and radiation know what to do, what to think about and how to keep me alive in my sweet dreams? These questions, atheists cannot answer. Unfortunately, I cannot answer them as well. I can only use my logic, common sense and what I know from modern science.

Among religious people, "enlightenment" sometimes can be replaced or coexist with a phenomenon of "vision." Scientists do not believe in "visions," considering this as a sign of a sick brain. However, if we accept the idea of signals from outside, this phenomenon might be both real and unreal at the same time. It is unreal in the sense that an image of Virgin Mary or Jesus Christ suddenly can appear in the sky and can be seen by anyone. But it is real in the sense that certain information, sent from the Universe and maybe not having anything to do with the image of the Virgin Mary, can find a corresponding brain-acceptor or part of it, that being excited by that radiation, can induce a quite real (for that particular person only) vision of the Virgin Mary. Both "enlightenment" and "vision" are just different forms of the same process - namely, the transfer of information and its acceptance by certain brain areas that are capable of reacting and being excited by waves of radiation sent by the Directing Power.

Some people explain "enlightenment" and "vision" by direct influence of spirits on our activities and thinking. Considering myself as an open-minded skeptic, who cannot blindly accept all paranormal claims, I did not want to include into this book those paranormal phenomena that are too weird to me at this time. Among them are: a soul materialization of dead people through ectoplasm; a visitation of survivors by materialized images of dead relatives; ghosts recorded on video camera; audio messages from dead people; TV images of the Other World and many others. I might include some of them in future editions of this book, if I have personal experiences with those phenomena.

Materialization of souls through ectoplasm released by a physical medium and their direct communication by voice and touch with live attendees is performed every week in Australia by David Thompson, who is considered the best of only four physical mediums on this planet (see his website at: www.silvercordcircle.com/davidthompson. html). These shows are performed for small groups and in full darkness

(Naturally!). Victor Zammit (retired lawyer) who strongly believes in a possibility of soul materialization, describes his own experience with this phenomenon in his book, "A Lawyer Presents the Case for the Afterlife" (see his website at: www.victorzammit.com). The main factor that allegedly makes soul materialization possible is the ectoplasm. Victor Zammit describes it as a mysterious fluid released from the body of the physical medium. In some other descriptions of this substance, on the Internet I could not find anything more scientific than the word "mysterious." Modern chemists, in cooperation with David Thompson, could easily come up with full descriptions of the ectoplasm to its very last atom. Alas, it has not happened. Naturally, there is no such cooperation since it would be the end of David's shows. I believe that his shows must be very exciting and well done, but presenting his shows as a real truth is, in my view, discreditable of his character. I also believe that my favorite magician – David Cooperfield – can do all kinds of materialization and dematerialization before larger audiences and in full light without any fraudulent claims and without ectoplasm.

Trying to find the answer to my question, "Who created the Mona Lisa?" I feel sometimes like I am looking for a bit of gold in a huge pile of stinky rubbish. It seems to me that the field of paranormal phenomena consists of about 95% frauds and 5% truth seekers. Such a ratio makes my search much more difficult, but I do not dismiss anything and even would like to see David Thompson's show and to shake hands with Jesus Christ or other celebrities who regularly attend his shows from the Other World.

Another weird paranormal phenomenon definitely deserves to be mentioned. This is audio- and video- messages from the spirits of dead people. If this would be a claim from a few paranormal mediums, it would not have much credibility. However, many thousands of people all over the World are now involved in this phenomenon and a huge accumulation of tape and video recordings have been produced to date. Such numbers do not allow outright dismissal of these phenomena unless it can be proved that it is a mass psychosis or mass deception. The official name for this is "Electronic Voice Phenomena" or EVP for audio messages and "Instrumental TransCommunication" or ITC for video images of spirits. In April 2008, Yahoo had 6,510,000 listings for EVP and 79,700 listings for ITC. Browsing through those numerous listings and reading the experiences of professional and amateur enthusiasts of this phenomenon, it is very difficult to figure out who is a fool, who is a crook and who is a serious researcher. However, after careful study of endless data of EVP and ITC, it is easy to conclude that a simple dismissal of it would be as stupid as to dismiss the phenomena of Kim Peek and Daniel Tammet described earlier.

The first voices from the Other World were captured on phonograph in 1938 and since that time have attracted lots of enthusiasts using all kinds of modern electronic equipment. There are even several software programs that help to purify those voices from background noise that is usually the main ground for skepticism. Hollywood could not ignore such great interest in this phenomenon and produced the movie, "White Noise." Only mainstream scientists and politicians still do not want to take it more seriously. When I think about it, I agree with their position, because it is much safer and more humane to keep crowds away from those facts that could cause a dangerous mass psychosis. However, such a position, as always, is just temporary.

At this point I do not want to write more about EVP and ITC until I have my own experiences. I will try to investigate these, as usually with an open mind, but with logical and healthy skepticism. If I am able to convince myself that this phenomena is not some kind of artifact (most of them have been already investigated), then I will present my views in the next edition of my book. Meanwhile, I will continue my search for truth and I would appreciate any help from my readers.

Quick test of your logic.

Dear Reader, I could give you at least several scientific facts and a few logical statements to convince you that we are not independent thinkers and that we are controlled by a force, which we know nothing about, yet. You might call this force "God" or "Intelligence Designer," but I prefer to call it "Directing Power." In order to make it simpler for you, I will give you just one scientific fact and just one logical statement. If you can argue against one or both of them, please, write me your arguments and I will either destroy them with my arguments or gladly accept them with an open mind. So, here they are:

1) **Scientific Fact:** Any move of body muscles starts with numerous signals of an electro-magnetic nature in the brain that are transformed into electrical current, running along neurons and causing very coordinated contractions of muscles.
2) **Logical Fact:** Something cannot be created from nothing.

Now, in order to prevent some of your arguments against these two statements, I would like to make a few comments about each of them. Brain signals of an electro-magnetic nature can be generated by two possible mechanisms: a) by biochemical reactions in neurons, b) by artificial excitation of neurons with electrodes, applying a force of an electro-magnetic nature. Both these mechanisms of signal generation are scientific facts. Can you argue against this? If not, let's go further.

Any chemical or biochemical reaction depends on laws of physics - particularly on the physical interaction of atoms and molecules. Such interaction depends entirely on two or maybe all three major forces: electrical, magnetic and possibly gravitational. The last force is too weak and I am not sure that it plays a significant role, but I just do not know. In simple chemical reactions, the final outcome depends just on those forces and on the concentration of reacting reagents. In more complex biochemical reactions some reagents do not participate directly in reactions, but create favorable conditions for reactions of other reagents. These are all simple scientific facts and you probably know them from your school years.

In a live cell, numerous complex macromolecules constantly react with each other under the control of genes, creating a sophisticated internal environment that is responsible for different cell functions. Each reaction of each molecule in cells must be controlled by the same natural forces mentioned above and by the same laws of physics. As I said earlier in the text of this book, these laws work by random chance. In other words, "A" can react with "B" only if they meet each other and have certain energy and conformation to create "AB" with new properties. If all molecules in our cells react on the basis of random chance, it would be absolutely ridiculous to assume that such random reactions could generate zillions of electro-magnetic signals in neurons that could generate strictly ordered movements of our muscles and transform these movements into music, painting, writing, etc.

Now let's go to our logical fact – "Something cannot be created from nothing." I hope you agree with this statement. Even if you tell me that radiation, which is "nothing" because it is immaterial, can become an electron, which is "something" because it is material matter, it would not be quite accurate. Radiation is still a piece of energy and cannot be considered as "nothing."

Now close your eyes to prevent visible light (electro-magnetic waves) from going through your eyes into your brain, since it can influence your further actions. And now make any unprovoked move of any part of your body. For example, raise your hand, turn your head, stick out your tongue, stretch your leg, or say something crazy. Let's assume that you turned your head. Nothing provoked you to do it. You did it just because you suddenly, with no reason, decided to turn your head. Here science tells us again that your brain has generated numerous signals that made the muscles of your neck contract in a very orderly pattern and as a result of such strictly coordinated contractions, your head turned to whatever direction you chose. The signals, that finally turned your head, are "something" of an electro-magnetic nature, but since your move was not provoked by anything at all, then you have to admit that "something" was created from "nothing," or you have to admit that this is completely against your logic and common sense. Would you agree with me that the overwhelming majority of our body movements (not all, of course) are not provoked by all those environmental factors (light, temperature, sound, etc.), that allow us to sustain all our physiological functions in compliance with dynamic changes of the surrounding media? If you agree with that, then how can we explain the generation of an infinite number of signals in our brain without any influence from any environmental factors from outside?

I am writing all these questions and speculations and my hand is dancing on the paper transferring my thoughts into letters and words. Strictly ordered moves of my fingers, my hand and my eyes are the results of signals from my brain. None of these signals is provoked by my environment. How and even why are they generated in my brain and making me sit with my computer, typing all this craziness, instead of giving me a chance to go outside and enjoy my life, fresh air and pretty women?

When Leonardo da Vinci was creating his Mona Lisa, every move of his brush was done by signals from his brain. If those signals are just physicochemical reactions in his brain, then how could all his proteins, carbohydrates, lipids, DNAs, RNAs and numerous other molecules be so smart to create such a masterpiece?

There is no answer to these questions. But if we continue believing that the chemistry of our brain is solely responsible for all masterpieces of human civilization, then we will never be able to figure out what Intellect is all about.

What is Intellect?

It is quite possible that the speculations in this book are nothing but a fantasy. I just tried to follow logical thinking based on available scientific information and follow my beliefs that our WISHES and THOUGHTS cannot be explained merely in terms of biochemical reactions in substrates of our brains. **There is something else that we still do not know much about; we can approach it only by logic and speculations.** My attempt for such an approach results in a conclusion about the possible existence of an Intellect in a different form of matter than the human brain, and about the constant exchange of information through radiation between this Supreme Intellect and the Human Intellect. This speculation might be completely wrong, but if we close our minds and reject any idea that is not in the line of evolution, we will stay trapped in the prison of our beliefs. Fish will never know that there is absolutely another world above the water. We know almost nothing about the world outside of our planet and virtually nothing beyond electrons, protons, quarks, photons and other particles or waves that are building materials of the Universe (and our Intellect). If we limit our quest for knowledge by merely our beliefs, then we will never know what is going on above the water.

It is very difficult to accept and to comprehend the speculations in this book. Rejecting them, we limit ourselves with only one alternative – namely, an independent work of our brain, based exclusively on the laws of physics and chemistry. Such an alternative is much more difficult to accept since in the core of these laws there is an element of pure chance. Besides, if numerous movements of our body and its peripherals are not induced by direct environmental signals (light, sound, heat, etc.) then such an alternative must assume that nothing can generate something (brain signals). Such an assumption would be completely void of logic and common sense. Although a signal of an electro-magnetic nature is not a material matter, it still cannot be generated from nothing, especially in a complex strict order to be transformed into a concerto or the text of this book.

On the other hand, if we accept the concept of a Directing Power, then its logical development leads us to the Sun as a main source of such Power. All live organisms of our planet are like sparkles in a big bonfire - born by it, flying in space and carrying its energy and then extinguishing, leaving just a tiny amount of ash. Sun is that big bonfire and we are all its "sparkles" carrying its energy for a limited time. After all, we consist mostly of water plus a small amount of solid matter, but we are distinguished from all inanimate objects by the fact that we carry energy from the Sun. The core of this energy is basically mutual transformations of protons, electrons and radiation, that are found exclusively in the Sun and in all live organisms, but never in inanimate objects unless we or the Sun create special conditions for releasing energy from inanimate objects (e.g. a nuclear blast).

If our Intellect is basically an energy emitted or absorbed by our brains, then why should we deny any Intellect outside of a brain in another form of matter if the same kind of energy emission and absorption is going on in the Sun and in billions of other stars, which together might represent the Supreme Intellect of the Universe?!

The view of the human brain, as just an Acceptor of information from outside, could probably be tested. I wish I could discuss this with physicists who work on the production of anti-matter. Maybe we could build a bubble surrounded by a very strong electro-magnetic field that could prevent any waves of any energy to penetrate such a bubble from outside. I wonder if prodigies with savant syndrome could show their unique abilities inside such a bubble. If such tests are realistic and can convincingly show that the brain operates absolutely independently of any influence from outside, then we should start paying more attention to another "no-no" of modern science – I mean the Soul. We know nothing about it and we do not want to believe in anything that is beyond our genes. We now have some technologies that allow us to go deeper into the chemistry of the brain. Will it help us understand how the Mona Lisa was created? Maybe – maybe not. The work of a Soul might depend on genes and chemistry, but I would be very surprised if all biological phenomena are just genes, chemistry and nothing more.

If we assume that the Sun somehow participates in the brain's work influencing our thoughts and decision-making, can we call it God? I think this would be wrong because the religious term "God" implies a certain Power that controls everyone and everything on the Earth, while the power of Supreme Intellect is not unlimited, but very influential. If God is all mighty and can control everything, then how can we explain numerous negative features of human civilization? How can we explain the Holocaust and many other crimes in human life? In order to keep the image of God clean, religious people created the opposite image – the Devil. I do not believe that there are two opposite forces like God and Devil and I am not going to discuss this here. The truth however is that good and evil are equal in this world. In my attempts to limit the discussion of this book to only one Directing Power, I would rather try to explain negative features of human civilization by different programs in our "Hard Drive," or by different signal acceptors in the brain that trigger certain actions, that might not be humanistic from the viewpoint of other human beings.

God's laws do not make the same difference between Good and Evil as we do in our perception of life. Imagine, for example, that you took a little beautiful Bambi and put him in the center of your backyard. Then you allowed two lions to enter your backyard. What would you expect the lions to do? Do you expect them to pet little Bambi and to sing him "Lullaby, Bambi...?" Of course, not. Bambi would be finished in a few minutes and the general scene of this event would remain in your mind as the ugliest example of an equality between Life and Death, between Good and Evil.

I do not like the hypocrisy of religious people, when they call God "Love" and often say even to a person suffering severely, "God loves you!" God's love has nothing to do with our perception of love. If it would not be so, we would have a perfect world. But our world is far from being perfect and most likely will never be. Nevertheless, we want to see God with our understanding of Love and Justice, but it is not easy. I do not believe that these qualities of human nature should be assigned to God's characteristics. The death of Bambi is a disaster from a human perception, but a natural event from God's viewpoint, whether we like it or not.

The power of Supreme Intelligence and its possible realization through solar radiation or through zero-point field energy might be limited and depends on the ability of an Acceptor (our brain) to catch and to digest information from outside. If we are just live computers, we need all the same things electronic computers need. First, we need "electro energy," which comes to us as solar energy, consumed through food and warm temperature, and second, corresponding "software" programs. Some of these programs we take from our parents in the form of a genotype. Some others we develop through our constant learning and we improve them during our life. Having a source of energy and programs, we, like regular computers, can make numerous decisions, which at first glance look completely independent. However, this illusory independence is not absolute, since any task, worked out by a computer, starts with commands sent by the operator's hand. In the described above unique examples of blind musicians or man-calculators we can see incredible work of the brain without any consciousness involved. These examples and many others provide clear indications of the Directing Power. Whether it comes from outside through solar radiation or it works from inside our brain, its influence should be limited by those programs that every live organism carries in its brain. These programs are somehow written in a substrate of our brain and can be divided into two general categories: inherited and acquired. Inherited programs (like DOS or Windows) we get from our parents, while acquired programs (like WordPerfect, PageMaker, etc.) we develop in the process of our growth.

The first category is influenced by evolutionary forces and probably weakly controlled by the Directing Power. The second category is influenced by social developments. Mozart, for example, would never become a great composer if he did not have a chance to be close to musical instruments. The control of a Directing Power in the development of these programs is more substantial, but completely depends on the ability of brain receptors to accept corresponding signals. I cannot exclude the possibility that an excitation of the same brain areas by a signal coded in waves could induce a work of different programs in our brain, and decisions or actions caused by these programs could have antisocial or cruel characteristics. Thus, the limited character of the influence from the Directing Power cannot be equaled to an all mighty God, who has too many features given him by people's imagination. Supreme Intelligence is not God (or at least is not God by a religious definition), but a certain quality of matter that is capable of not just holding energy inside, like uranium, but being with it in a constant mobile equilibration coming from energy and into energy.

If such a fantasy has at least a little bit of sense, then it still remains unclear how mutual transactions of protons, electrons and radiation could create a quality of the Supreme Intelligence or our Brain Intellect. On the other hand, we accept without questioning the fact that material particles can be turned into immaterial radiation and vise versa. We take this as a scientific fact, although we are absolutely unable to comprehend such transactions, as we are unable to comprehend at this point a work of our own Intellect realized through three forms of matter. This is probably the main secret of Mother Nature and I am not sure that it will ever be revealed.

Is Intellect a blessing or a disaster?

Many years ago, the Devil made Eve disobey God and take an apple from a forbidden tree. (Oh! Women!). This apple gave her and her poor boyfriend a distinctive quality that we now call "Intellect." Whether it was that way or another, the moment when Man got an intellect has become the beginning of his end, and the saddest event in the history of our planet. From that moment, Man received a power that put him above natural laws of a biosphere. He started to change it actively and irreversibly. In the beginning, this process was going slowly due to the shortage of "know-how" and "man-power." However, the speed of this process (which we mistakenly call "Progress," but which in reality is "Regress," destroying the biosphere and degrading Man) was continually increasing. Each new twist in that direction improved the life style of human beings, decreased their death rate and inevitably accelerated the population of intellectual creatures on our planet. The Intellect knocked out from Nature its ability to regulate the development of the biosphere and to maintain a balance. Those who understand human nature will agree with me that Intellect is a time bomb, which sooner or later will inevitably explode and destroy itself and everything around it.

Many educated people warned us about the problem of overpopulation a long time ago, but at that time it wasn't so visible or so serious. Even now many people, who are not able to see beyond their noses, do not believe in this problem. And those who can see, are afraid to talk about it or just do not care. Many anti-abortion and religious groups strongly believe that it is so inhumane to call for the regulation of the human population! Many others argue that the problem is not overpopulation, but a human greediness that leads to hyper industrialization of the entire planet leading to its own destruction. I will address this problem later. At this point however, I want to remind you that both of these problems are secondary. The primary problem is Intellect.

If a human being would remain an equal member of the animal society (to which we all belong), then Nature would be able to preserve the balance on this planet. However, the ageless war between Nature and Intellect is still not finished. What will happen then in the next 100-300 years? In the next 100 years, the total human population of this planet will be increased from 6.5 billion to 16 billion. Only a complete idiot can believe that it is realistic to accommodate, feed and employ all these people. Problems of food and energy will become the most vital. These problems can be observed right now in those spots of our planet where the reproduction rate is especially high. In this respect, African and Asian countries are in the front line. In Africa, for example, every woman has an average of eight children. Taking into account that large areas in Africa are not suitable for agriculture, the shortage of food is already quite alarming. This situation immediately results in local tribal wars, where the political game is just a fig leaf, masking the simple and violent struggle for survival. Wars in turn lead to the destruction of the economy and to famine claiming thousands of lives.

The problems of food and energy are progressing hand in hand aggravating each other. For example, in order to stop the movement of desert to agricultural fields, governments spend a lot of money planting trees to stop ground erosion. However, people need a source of energy for cooking and many other purposes. The result is that everything that can burn goes into a bonfire; soil is eroded and becomes unsuitable for agriculture. If soil cannot provide enough food for a dense population of Africans, they have no choice but to seek food among wild animals. As a result, right now a mass hunt is going on in virtually all national parks and the only purpose of this hunt is to feed people. Can "Greenpeace" or any other organization stop this process? Of course not! And it is not so difficult to foresee what will happen with unique African fauna in the immediate 100 years if no one takes seriously the problem of demographic explosions. It is also not so difficult to foresee those incredible sufferings to which people willingly doom themselves using their "Intellect" that destroys them and everything around them.

Now, to the joy of "Pro-Life" activists we have efficient fertility drugs on the market, which soon can be bought "over the counter." One young couple already had three children, but she wanted twins. She took fertility drugs and delivered quadruplets. Now they have seven children, but she wants more. Why not? Especially, if taxpayers are going to pay for all the expenses. Looking at them on TV, I felt remorse for the guy, who found himself in a trap of a baby-machine. At the same time I could not understand the burning desire of his wife to give birth to more and more babies. It is a very irresponsible and selfish attitude, but there is no doubt that fertility drugs will be of great demand in the market. (The record in 1997 is seven babies delivered from one pregnancy by one woman after a course of fertility drugs. Who wants more?) (The record in 1999 is eight babies! Oh! God! Stop this madness!)

Is overpopulation just an African problem? Of course not. Go to Mexico, Delhi, Jakarta, Caracas, Cairo and many other cities and you will see results of overpopulation and overindustrialization. You will see what polluted air people breathe, what terrible food they eat, what terrible water they drink. Looking at them, I cannot understand the position of the Vatican that still cannot rid itself of dogmas that forbid abortions and even condoms. Having an enormous impact on the majority of people on this planet, the Vatican deliberately makes their suffering worse and worse every year, making a miserable and very unhealthy life for millions of people. I am sure that one day the Vatican will be forced to remove its shortsighted dogmatism against abortion and birth control; however it must be done now, to prevent the consequences of an unbelievable catastrophe.

Is it possible that the same problems will be observed in "the Americas" and European countries? No doubt about it! In fact, they are here already. According to the latest statistical data from the American Census Bureau, the population of the United States will be increased by 50.2% in 60 years and approach 383 million in 2050 year. It is not a surprise that white people will be increased by 29.4%, while blacks - by 93.8%, Asians by 412.5%, Indians and Eskimos by 109.1% and Latinos by 237.5%. This growth is not only due to high birth rates, but also because of legal and illegal immigration currently approaching about 1.5 million a year.

At the time this chapter was being written, people of Haiti started to build all kinds of simple sea vessels in order to start mass immigration to USA shores as soon as Clinton would come into power. Why? It is very simple - no one enjoys hunger. But maybe they are hungry because their president was disposed by a military regime? NONSENSE! The real reason is the same - overpopulation. By the same reason, thousands of Mexicans and other Latin-Americans illegally cross US borders every month. The scale of illegal immigration will be increasing and nothing, absolutely nothing, can stop this process. And any politician will never openly and honestly discuss the real problem and how to cope with it. As a result, in our "God-blessed America" chaos will prevail sooner or later. Huge masses of jobless, bums, the sick, drug addicts, prostitutes and criminals will inundate all American streets, undermining economical and social principals. This process exists already and not only here, but in European countries as well. Internal problems and furious struggles with them will lead to serious international conflicts and then to an inevitable nuclear resolution.

Walk along the streets of New York, Chicago, Detroit. You can meet many tramps, beggars, addicts. People here look nervous, stressed, preoccupied, often rough and unfriendly. And after you get a clear picture, take a trip to a remote village in any place in the USA or abroad. There, you will not see beggars, addicts or prostitutes; and people there will be friendly and affable. They appear much healthier than people in large cities, because they work in open air and eat natural food without any chemical additives. Should we all try to follow this life style? As long ago as in the 18 century the French philosopher Jean Jacques Rousseau exclaimed: "Back! To Nature!" He, like many other intelligent people before or after him, perfectly understood the direction in which human society is going, and the problems Intellect can create. He also understood that his appeal would not be heard and would not be taken seriously. To go back to Nature is as unrealistic as to reverse time. It is interesting that his appeal was made when overpopulation was not yet a serious problem and only social injustice prevailed.

Curiously, among "intellectual" creatures there is a small percent of human beings who have a better perception of common sense. However, the rest of the World calls them "wild" or "uncivilized." One example is the Indians of the Amazon River, who contrived to preserve their lifestyle in close connection to Nature and for many thousands of years. Another example is even more amazing. In the heart of a highly industrialized America, there are people like us but they call themselves Amish and Mennonites. They live for centuries in the World that we mistakenly call "civilized." However, they contrived to separate themselves from a technological progress and to preserve close connections to Nature. They do not use electricity, do not watch foolish TV programs, do not drive cars, use only manual tools, cook only natural products and drink only natural water, milk or juice. And they do not have a sexual dissoluteness, which "civilized" people call "sexual revolution." This lifestyle is rejected by coddled people or rather spoiled and weakened by civilization. They cannot imagine their life without TV, cars, air-condition, Coca-Cola and many other "toys." But they easily replace books with comics - Long Live Civilization!!!

The above examples show that Intellect can actually be sensible and reasonable, but for some reason it can be seen only in a very small number of people.

The second problem, which now becomes more visible than in the past, is the problem of unlimited consumption. However, this problem is usually discussed separately from the problem of overpopulation when both of them are closely connected. The problem of excessive consumption cannot be resolved by an abstention of consumption. Our government constantly is seeking new ways to have people consume more products and goods. All efforts toward increased consumption can take a country out of recession, can stimulate the economy, can increase employment and by this can employ and feed up all those numerous coddled creatures who constantly demand from the government more decisive actions in order to let everyone consume, consume, consume. In order to calm down all these creatures, the government is trying to stimulate the economy by building new plants, factories, cutting more trees for lumber, taking more fertile land for construction, and at the same time ignoring (no time to think) the destruction of the ozone layer, global warming, pollution of rivers and oceans, and even harmfulness of many consumable products. This results in a "Catch-22." People develop more consumable products that affect their lives and promote the growth of population. Increased population promotes further growth of industry in order to let people consume more products. Thus, both processes are closely connected, influencing each other and always halting with a short recession when we make more products than we can consume.

How far can these processes go? It is already clear that it is about time to stop, if we do not want to doom future generations with incredible sufferings. But how can we do it? If we stop the growth of industry at the current level, immediately economic depression will affect the lives of millions of people. Many of them would lose their jobs, crime will be increased and life will become miserable. And immediately angry crowds will start marching and demanding from their leaders certain actions to correct this situation. In order to prevent such a situation, the economy must always go up and the national product must always increase. But how can it go up all the time? It cannot, but it cannot stop either. This is an irresolvable puzzle. The human brain was constructed in such way that it cannot stop the destruction of this planet and itself by itself. It cannot come out of a Catch-22 without help. But this help can be given to the Intellect only by Nature or by God, and the only realistic application of such help would be a drastic decrease of the human population.

Another destructive effect of Intellect on human population is quite paradoxical since it is in direct proportion to medical progress that is certainly the most desirable, but a very dangerous product of Intellect. How does it work?

Mother-Nature does not like mistakes. It destroys all living creatures, including human beings, that acquire certain errors in programs of their development. Scientists call such errors - "mutations" and they usually happen in an occasional substitution of chemical groups in the most important chemical substrate where virtually the entire life of an organism is coded in so called - Deoxyribonucleic Acid (DNA). Errors or mutations can happen very early in sperm and eggs or much later when fusion of the sperm with egg send the most important signal to DNA letting it know that it can start decoding the development of a new living creature. If the error is very serious, then such a development can be terminated right in the beginning. Mother-Nature wants only normal children who are fully capable to compete for their lives. However, if an error is not ultimately dreadful, a new live organism can be born and it might have either advantage or disadvantage in survival as a result of that error. An advantage is always welcome by Nature, while errors carrying a disadvantage in a struggle for survival do not have any mercy; they must be eliminated to keep the pool of genes healthy and strong for the continuation of life and survival of the species.

If Nature would be more merciful and a live creature with bad genes could easily approach the age of maturity, those bad genes would be passed to the next generation and then again to the following one. Generation after generation bad genes would be accumulated in a genotype of such an unfortunate species making them weaker and weaker until the entire population would turn into sick and crippled individuals doomed to final extinction. This does not happen in all species, except just one - the human race. In the animal world natural selection is a very efficient tool for keeping the pool of genes clean from any undesirable errors. Those who were less fortunate in having their genotype as good by all standards, do not have a chance to survive in the long run even if their bad genes can be passed to a few following generations.

In the human world it does not work and natural selection is going in reverse by preserving as many bad genes as possible. Intellect gave us enormous advantages for survival and put us above the animal world, but as I said earlier, in reality it is just a time bomb.

We have made remarkable progress in medicine, can explain now numerous disorders, are able to cure many diseases and can easily save the majority of bad genes that otherwise Mother-Nature would destroy without hesitation. Our Intellect has created endless numbers of social and ethical laws along with numerous medical approaches that allow individuals carrying bad genes to achieve their age of maturity and pass those genes to their children. Certainly that from a humane viewpoint it is the right thing to do. However, from the viewpoint of evolution, an accumulation and nurturing of bad genes leads species to weakening and extinction.

Since the human race is the only species where Intellect is in constant struggle with the laws of Nature, the weakening of the human race is a very natural, logical and inevitable chain of events. This fact imposes a significant burden on social institutions of the human population. Our Intellect tells us that the nurturing of bad genes is the right thing to do and we do it since an alternative way would be extremely inhumane. We are proud of our humanism and feel that we are better than all other live creatures on this planet. Surprisingly nobody wants to ask the question of whether it is really humane to accumulate bad genes and by so doing to increase the number of individuals suffering from genetic disorders resulting from these bad genes. We do not want to think about it. We are too busy in search of other means that could help human beings to enjoy their lives regardless of their medical problems. We are led by Intellect, rely on it and are not able to separate ourselves from it.

One example of our humanistic nature is especially bright. She is mentally retarded and her muscular dysfunctions bind her permanently to a wheel chair. He is also mentally retarded and has genetically caused bone problems that made him severely crippled. However, their sexual organs did not have any problems and they decide to get married. Needless to say both of them live under full time social care and both of them are physically and mentally incapable to raise children. Do they have a right to have children and pass their bad genes to the children's genotype? Who is going to raise their children? Of coarse, they have a right and of coarse, we (society) would take full care of their children. Oh! We are so good, so humane!

A Reader might hear a bit of sarcasm in my words. I want to emphasize that I am proud of the humanistic nature of human beings and certainly I would not support any idea of physical extermination of individuals carrying bad genes. The only purpose of this chapter is to show that from a biological viewpoint Intellect is not God's gift, but a tool of self-destruction. The means of self-destruction are coming from different directions, but Intellect is responsible for all of them. The accumulation of bad genes is one of such means, but it probably will not cause extinction of the human race or at least will not be the major force.

Inability of the Intellect to put under strict control overpopulation and overindustrialization is a more serious destructive factor, which can be visible right now, long before the accumulation of bad genes can make a significant impact. On the other hand, the burden on social institutions mentioned earlier is already quite visible. Numerous cases of cancer, autoimmune diseases, mental dysfunctions, genetic abnormalities and endless numbers of other disorders are just a result of the accumulation of bad genes in the human genotype. At this point every one of us is carrying numerous bad genes. Not all of them will be able to affect our health or our abilities due to complex interactions of all genes. However, we pass them to future generations and sooner or later they will show up bringing suffering and despair.

Is there a way to turn our Intellect into Mother-Nature's ally instead of her enemy? No, this is not realistic at all, since Intellect categorically refuses to live under the laws of Nature where death and natural selection resolve virtually all problems. Realizing that Intellect is a self-destructing power, it is easy to see that in the long struggle between Nature and Intellect, Nature will finally win, and Intellect will commit suicide. But before it happens, this Planet will see too much suffering and too much destruction, and absolutely nothing can stop it.

Even Masters-Gods from Dr. Weiss's book (see Chapter "Soul, How real is it?") agree with me saying: *"Everything must be balanced. Nature is balanced. The beasts live in harmony. Humans have not learned to do that. They continue to destroy themselves. There is no harmony, no plan to what they do. It is so different in nature. Nature is balanced. Nature is energy and life... and restoration. And humans just destroy. They destroy nature. They destroy other humans. They will eventually destroy themselves. It will happen sooner than they think. Nature will survive. Plants will survive. But humans will not..."*

This is a pretty gloomy prediction that in my view is absolutely correct. However, if the Masters-Gods are powerful and influential in all events on this planet, then I would like to ask them a simple question: "Why the Hell did you give intellect to humans?"

What is the most important thing of everything?

I am looking through my window at the big tree friendly waving to me by its branches. Nine squirrels are chasing each other, jumping from branch to branch with amazing skill. They are God's creations, as we are, but they do not think about God, they do not pray to God, they do not go to a church. Such things are not important for them. They only know that today is a sunny warm day, there is plenty of food around and life is just wonderful. Why then cannot human beings enjoy life, like squirrels do? Why do we have to think about God, matter, energy and how it all works together? Well, we do not have to, but whether through Eve's disobedience (God knew what he was doing) or by some unfortunate gene mutations, we got that weird quality called Intellect. This quality makes me now to write this book, to think about God, life, consciousness and many other things instead of going outside and enjoying a warm sunny day, like squirrels do. Is it important to know if there is a God or if there is no God? What is the most important thing in the life of any live creature? Let's try to find the answer at least to this simpler question.

This question would never be raised if God did not create Life on this planet and if he did not make a terrible mistake giving an Intellect to one of his children. But what has been done, is done and now the one intellectual creature of this planet is trying to figure out what the most important thing is.

First of all, this philosophical question has to be discussed inside of a certain time frame, since the importance of things can be quite different if time is just a momentum, or if it goes to infinity. For example, at this particular moment I have to run to pick up my son from summer camp and this is the most important thing of everything for me. On the other hand, if time is infinity than the importance of things loses its sense since it makes sense only in close conjunction to life.

Thus, the discussion of the above question must be done only in a time frame of an individual life and never outside of life. Besides, it can be discussed only in conjunction with the life of an intellectual creature, that is a human being. Any other live species never asks themselves any philosophical questions. For them it is not important at all.

Now, if we agreed with the boundaries defined above, can we say that Life is the most important thing of everything? If we say so, we have to admit that Death is an equally important thing, since a Life of one creature fully depends on the Death of other creatures. Life and Death are not separable. They are Yin and Yang defined by Taoists many centuries ago. The well known symbol of Taoism shows Yin in Yang and Yang in Yin symbolizing polarities in unity and harmony.

Nobody would say that Death is the most important thing of everything. But if so, we cannot say the same about Life as well, since they are also polarities in unity.

Our life on this planet is feasible only because of the Sun, which gives us the most important factors for life - Light and Temperature. Can we say then that the Sun is the most important thing of everything? No, we cannot. First, there are billions of billions of other suns that do the same thing in other galaxies and they are not and cannot be important to us except for some poets who can get an inspiration only looking at bright stars in dark skies. This means that our Sun is very important to us only, but not "an important thing of everything."

Second, our Sun is not forever and it will die in about five billion years from now leaving our planet cold and empty or swallowing it due to the constant approach of the Earth to the Sun. One way or another, Life on this planet will be ended and the importance of the Sun will be erased from the records of the Universe.

As I said above, an importance can be discussed only in conjunction with our consciousness. If I am dead, nothing can be important to me. Also, if I am in a deep sleep, my brain does not care what is or is not important as long as my body feels comfortable in bed and gets enough oxygen from the air.

Thus, only the state of awakened consciousness of our Intellect can evaluate the importance of things around. Here comes an interesting sophism: The most important thing of everything does not exist out of our Intellect. This means that our Intellect generates the very essence of important things. When Intellect disappears (e.g. by death), importance disappears too. Hence, Intellect and importance are virtually the same thing, since they cannot exist without each other, like Yin and Yang. Does this mean that the most important thing of everything is our Intellect, our consciousness? Basically this is true, but a little later I would like to give another answer to the question of this chapter.

The life of every person is the result of a very, very long chain of coincidental events that are not even possible to evaluate mathematically. If 1000 or 200,000 years ago Jack would not meet Jill, I might not be here today writing this book. During those long years unbelievable numbers of coincidental events were responsible for my existence today. Even considering the time after the marriage of my parents, the probability of my appearance on this planet was one in many billions. Since my life is such an extremely rare combination of numerous events and never ever will be given to me again (unless my Soul reincarnates into a skunk or hyena), I have to see it as a wonderful gift given to me by God or by circumstances to appreciate this divine miracle in its full extent.

Unfortunately, the ability to appreciate this miracle comes to us usually too late. We waste most of our years for fulfillment of our futile and unnecessary egos resulting from a constant brainwash by society. Our Intellect tricks us and makes us forget what is the "most important thing of everything." Only much later in our life do we start realizing that most of it we put into a garbage can and is gone forever.

The majority of people on this planet cannot enjoy life to its full extent. They are slaves of wrong ideas or habits, greediness or hypertrophied egos. They are slaves of their own judicial system, slaves of numerous crooks in business and politics, slaves of material things and of many other factors.

This slavery is designed for one clear goal - to give the human species an advantage in survival by making all individuals work for the benefit of mankind. Our Intellect is fully responsible for this slavery, but the benefit to mankind is just an illusion, since we work for our own survival, but we destroy mankind, our Intellect and our planet. There are also such things as Obligations and Morality, which also impose certain restrictions on our lives. We must take care of our elderly parents, who sacrificed many years of their lives to give us whatever we needed to appreciate our lives. We also have to accept restrictions of morality. Although morality is a personal code and it can be quite different among people, but whatever morality we accept, it gives us certain restrictions in our perpetual conquest to enjoy our lives.

However, as usually, where Yin is - there is Yang too. Our Intellect allows us to get full appreciation of our life, our existence. It gives us a broad set of emotions that make our lives even more beautiful. Still, realizing the self-destructive force of Intellect, I cannot say that it is the most important thing of everything. Instead, I would like finally to give my own answer to this question. Here it is:

Since an individual life is such an incredibly rare result of coincidental events and is very short, it must not be wasted by futile and unnecessary activities. The most important thing of everything is to enjoy your life as much as you can in the frame of your obligations, your morality and your taste. If you enjoy watching the sunset, go to see it more often. If you enjoy listening to music, do it more often. If you enjoy being around kids, make it possible. If you enjoy watching TV shows for idiots, like Capital Wrestling, watch them as much as you can. If you enjoy drugs or cigarettes, take them - it is YOUR life and you have to enjoy it in the frame of your desires, even if you are stupid enough to make it much shorter. When life is over, nothing has any sense of importance in the entire Universe. Life is a wonderful gift, and enjoying it is the most important thing of everything.

Dear Reader, I do not expect you to approve or disapprove of the viewpoints and speculations in this book. You have to take it just as a subjective, individual way of thinking that may or may not be in accordance with your personal way of thinking. However, no matter how much you agree or disagree with this book, I would be happy to hear your comments about it, your own viewpoints or your objections. Please, send your comments to: intart@peoplepc.com

Thank you.

www.ingramcontent.com/pod-product-compliance
Lightning Source LLC
Chambersburg PA
CBHW030904180526
45163CB00004B/1700